THE END IS NIGH

The End is Nigh

A History of Natural Disasters

Henrik Svensen

REAKTION BOOKS

For my kids, Oleander and August

Published by
Reaktion Books Ltd
33 Great Sutton Street
London ECIV ODX, UK
www.reaktionbooks.co.uk

This book was first published in 2006 by H. Aschehoug & Co. (W. Nygaard),
Oslo, under the title *Enden er Nær* by Henrik Svensen
© H. Aschehoug & Co. (W. Nygaard) AS

English-language translation © Reaktion Books 2009

This translation has been published with the support of NORLA

English translation by John Irons

Printed and bound in Great Britain
by MPG Books Ltd, Bodmin, Cornwall

British Library Cataloguing in Publication Data

Svensen, Henrik, 1970–
The end is nigh: a history of natural disasters
1. Natural disasters – History
I. Title
904.5

ISBN: 978 1 86189 433 5

Contents

When Climate Becomes Disaster

With the new reports from the UN Climate Panel in 2007 came confirmation of something many people have long suspected: we are living in an age where climate change is a reality. A large number of people will probably thus be affected by more extreme weather and natural hazards linked to floods, landslides and storms. Since 1970 almost five billion people have been hit by natural disasters. Over two million people have lost their lives.[1] Behind these figures lies information of great relevance to our future. The disasters that affected most people in 2006 – and in the entire period since 1970 – were linked to too much or too little water and wind: drought, floods and hurricanes/tornadoes. In 2006 no less than 92 per cent of natural disasters were caused by dangers linked to various kinds of extreme weather. It is precisely these types of dangers that we are aggravating by our emissions of greenhouse gases. All three elements that affect how vulnerable we are to the forces of nature – exposure, resistance and ability to reconstruct – indicate that the more than a billion people who live in poverty will be worst hit. Even if one compares them with social disasters, natural disasters are commoner than most people think, and cause major problems throughout the world. As the UN's former relief co-ordinator Jan Egelund pointed out in 2007, natural disasters affect seven times as many people as wars do.[2] Recognition of the fact that natural disasters represent profound crises is now taking place.

The key task in the years ahead will be to find out how one can best understand the consequences of global warming – with regard to both individuals and societies. Computer simulations can prepare us for a certain number of degrees of warming and for how much the sea level will rise. Changes to precipitation patterns indicate where floods, periods of drought and landslides will occur. But how can we gain knowledge about the ways in which people will react, and if the climatic changes will result in cultural or social changes? We can make use of our knowledge of natural disasters. All the changes that will come in the wake of climatic changes have already been experienced by thousands of local communities already hit by natural disasters. For that reason, a retrospective look at the consequences of natural disasters is more important than ever before.

The major natural disasters of recent years in Iran (2003), Southeast Asia (2004), the USA (2005), Pakistan (2005), Burma and China (2008) have turned the lives of an incredible number of people upside-down. But have these disasters led to social changes? What effects have natural disasters had on the people who lived through them – and how are they explained? What is the prime cause of natural disasters, the force of nature or the insufficiency of the societies to respond to them? This book investigates why natural disasters occur and what they do to us. It explores the best-known natural disasters and their consequences for people, societies and cultures. I shall also take a more in-depth look at natural disasters that are seldom touched on by the media. Some of the disasters are among the worst events that ever took place in human history.

There is no reason to believe that the number of natural hazards will soon start to decrease, nor that we will witness a radical change in the vulnerability of societies. This means that the social causes of natural disasters will be even more obvious in the years to come, as the climate changes. Climate refugees are already a reality, and even a small change to the level of the sea can create tens of millions of refugees in low-lying countries. Despite this, discussions about the consequences of disasters for social organization and political stability are rare in climatic contexts.

Understanding of natural disasters, their causes and consequences is continually growing. A number of fields of research are now contributing with dedicated studies of disasters. In order to exploit the breadth of today's universe of knowledge, the approaches in this book range from geology to geography and history, via anthropology, sociology and the history of religion.

By focusing on natural disasters through historical time, as well as in various cultures, it is possible to investigate whether there are basic patterns to how human beings react during crises – and how the crises can be overcome. This book illustrates the history of natural disasters so as to prepare us for what lies ahead.

INTRODUCTION:

On the Edge of the Sahara

Traffic in the city centre was virtually at a standstill. It had been re-routed to protect the American embassy after the terror attacks in the USA in September 2001, and the chauffeur was visibly irritated. A large sticker with Bin Laden's lugubrious face adorned the back door of the green minibus butting its way ahead of us. It was Monday morning in Mali's capital, Bamako, and we were on our way to a meeting with the Energy Minister. Myriad market vendors lining the street sold clothes, vegetables and body parts from exotic animals (dissected monkeys and lizards) for ritual purposes. Bamako is famous for its fashions – and both men and women wore highly-coloured yellow, orange, red and green ankle-length costumes.

We eventually arrived at the department building, and had to wait for the minister in a dark corridor on the second floor. VIPS are normally late. Finally, we were shown into an anteroom, where we waited for a while before being led into the meeting room. The walls were decorated with portrait photographs of men in suits. The leather chair at the end of the two long tables was empty. After a further wait, the minister came in. He was powerfully built, with short hair and well-trimmed sideburns, and wearing a light-green costume. We got up. I was a little embarrassed, as I had forgotten to bring along any clothes for ceremonial occasions. When the greetings and formalities were over, the minister asked for a briefing as to what our plans were. He was worried about a volcanic eruption near Timbuktu and afraid of a possible large-scale evacuation. We put forward a plan of action, and he listened attentively while slowly twining his fingers. The minister concluded the meeting by saying that he would send one of his own observers along with us. The president had also been informed.

The aim of the expedition was to look for possible volcanic activity in the flat and sandy part of the country near Timbuktu. Since the 1960s French researchers had suspected that the earth's crust in this part of Mali was beginning to crack. Unexpected earthquakes in the 1990s seemed to confirm that a volcanic eruption was a real threat. We had received reports in autumn 2001 that this activity was increasing and was threatening a number of villages in the north of the country. Foul-smelling smoke was pouring out of holes in the ground, with temperatures of something like 400°C. In the most exposed area the village of M'Bouna, with 1,600 inhabitants, lay in the danger zone. The government was afraid that thousands of people could very soon be affected. Even though the area did not have any volcanic mountains, the observations were typical of activity one finds close to volcanoes, where warm springs and gas leakages are common. Volcanoes can begin where the ground is perfectly flat – and I recalled the story of the farmer in Mexico who witnessed the birth of a volcano in his cornfield in 1943. One year later, it was 335 metres high, and the lava flows had spread over an area of 25 square kilometres.

In December 2001 an expedition was planned in Oslo consisting of two geologists (including myself), a physicist and a film photographer. The Geological Investigation in Mali was to deal with the logistical aspect of the trip.

The trip from Bamako to Timbuktu was in a small propeller-driven aircraft. Outside the airport, two 4WD Toyotas were waiting. The airport lies on the outskirts of the town, and the road was full of sand, with scattered buildings on either side. Signs belonging to Western relief organizations had been placed at various points along the road. Drought and civil war between the settled inhabitants and the nomads during the 1990s had made the need of relief great. Diverse projects were being implemented in Timbuktu, such as one to help child-soldiers back to a normal life, and one to digitalize manuscripts from the once so famous university in the city.

Conditions in Timbuktu are not what they once were. Not so long ago, the city was one of the most inaccessible places in the world, fabled for its riches. It was not until 1828 that the first European managed to enter the city and subsequently return to Europe alive.[1] Now, you can surf the Internet from cafés like the one only a stone's throw from where we were sitting and eating chicken with sand in the sauce.

From Timbuktu we drove westwards through the arid, sandy landscape towards Lac Fabuibine and M'Bouna. Large, white sand dunes on the outskirts of M'Bouna reminded us that the desert can take over at any time. The narrow village streets were covered with sand, and in the main street there was much lively trading, with stalls where jewellery, meat and dry commodities were displayed on carpets under the low winter sun. There were no other cars than ours to be seen here. A gang of teenagers ran after our cars when we arrived. All the houses in the village were of clay, and we were to live in one of the few that had two floors, one with a power supply unit.

After a couple of days' fieldwork, we discovered that the incredible reports we had received in Oslo were correct. During our first day in the field we found that the landscape was dominated by scrub that was partially destroyed by fire and smouldering ground. Some of the holes glowed with a temperature approaching 800°C, which was much hotter than we had heard in advance. You had to watch your step – the ground underfoot was loose and you could easily trample through it. Elsewhere, there were open fissures several metres long, with sulphur and red-fired clay along the edges. It seemed obvious that something was going on under the surface that, at worst, could lead to a volcanic eruption.

The inhabitants of M'Bouma were very worried. During a conversation with the mayor it was stressed that they feared the phenomenon could develop into something disastrous. Already, they were feeling the effects. The grazing areas had been ruined, and the smouldering was getting closer and closer to the village. Occasionally, the distinctive but indefinable smell the wind brought with it could be made out. The mayor was uneasy. Furthermore, the quality of the water had deteriorated, he explained. The village lacked deep boreholes. And the young people were leaving. The future did not look bright.

The fieldwork was short and intense. There was a lot to look at, with long stages by car in a landscape without any roads. On

Thursday evening, two days after our arrival, it was already time to pack our things. We had to get back to Timbuktu the following day in good time, before it became dark, in oder to catch the plane back to Bamako. The chauffeurs were also armed.

From the terrace, there was a view out across the rooftops. A group of camels was resting on the top of a sand dune in the distance. The four of us discussed the day's findings with the other geologists from Bamako. The minister's delegate was sitting in a chair, crouched over a transistor listening to a direct football transmission from the African Cup, which was being organized in Mali. He would rather have been back in Bamako watching it on TV, but all the time he had kept an eye on what we were doing, even if at quite a distance. Our two chauffeurs were making tea over a small cooker and serving everyone in turn. Then they took out their prayer mats and faced Mecca. There was also a delegate from the High Commissioner in Timbuktu on the expedition, and the local guide – who happened to be the brother of the mayor in M'Bouna and was easily recognizable by his bright-blue turban. Dinner was carried up from the backyard and consisted of a chicken hotpot with rice and bread. There was the slight crunch of sand between our teeth throughout the meal.

After having discussed the fieldwork, we arrived at the conclusion that there were a number of observations that did not agree with a volcano theory. The scorched areas proved to be circular, something that indicated a starting point at the centre. This accorded badly with ignition from magma, and we had not found any signs of lava. Could it be that there were completely different causes for the phenomenon than the beginnings of volcanic activity? There was still a little time left, so the last day was spent testing out a new hypothesis.

Out on the terrace in the darkness of the evening, I was probably not alone in thinking that time had passed a bit too quickly and that I was not yet ready to leave M'Bouna. Life in the village did not seem any less exotic because of the stream of music that stole over the rooftops. A couple of us went down into the backyard and out into the street to track it down. It came from the outskirts of the village. The road was full of sand and heavy going, but at least clearly visible in the bright moonlight. The houses cast long shadows, and we rounded the corner of the last house in the village. Through a fence we could make out a lit backyard with some young people. We moved further along the fence to get a better look, but were soon spotted. A couple of them came over the gate and invited us inside. In the poorly lit yard were about ten youths sitting in a circle. Inside the ring, about just as many girls were dancing to pop music from a cassette player. This was the village 'discotheque' and meeting place. One of the major problems in M'Bouna is young people leaving, so this gathering probably had the mayor's blessing.

What would happen to the village if a volcano were suddenly to erupt in the vicinity? Would the population have to leave M'Bouna for some refugee camp in the vicinity of Timbuktu, or even farther away? All of us knew that our results could be crucial for the future of the people in the area.

DISASTERS AND US

The ways in which people and societies protect themselves against and react to natural disasters says a great deal about the values and priorities of societies. Natural disasters are consequences of encounters between nature and society, the extent of which depends on how well people have managed to adapt to dangers. Furthermore, disasters are something that belongs to everyday life. On average, between one and two natural disasters occur on the earth every day – and natural hazards surround us at all times. So nothing extraordinary is required for a natural disaster to take place. Such disasters do so continually, without our necessarily reflecting on how they affect us. One of the aims of this book is to show how over time natural disasters have left traces behind in us and in what ways our ideas about society and nature alter during times of crisis. First of all, we have to know why natural disasters occur, and what they are.

Natural disasters occur when a vulnerable society is exposed to a natural danger – a 'hazard'. This hazard may be an earthquake or a violent volcanic eruption, or longer-lasting events such as floods and drought. Disasters can be classified according to the triggering hazard, and the main hazard types are geophysical (earthquakes, volcanic eruptions), hydrological (floods, slides), climatological (extreme temperatures, drought, wildfires), meteorological (storms), and biological (epidemics, insect infestations). Drought and epidemics often have a stronger social cause than do geophysical disasters, although they too are triggered by a natural hazard and a society's lack of ability to deal with changes in the environment.[2]

Ideally, extreme natural phenomena do not lead to natural disasters if societies are invulnerable. But all societies are vulnerable to various degrees, depending on such factors as economy, politics, religion and their social structure and infrastructure. Furthermore, the distribution of vulnerability in many societies is uneven. One's position on the social ladder is often the most important factor. The extent of natural disasters can be defined on the basis of various criteria. They can be divided according to the size of the affected area, the number killed, what these deaths were caused by, the societal consequences or the economic losses.

To what extent are societies influenced by natural disasters? Obviously, large natural disasters lead to material destruction on a huge scale as well as economic burdens. But, as with all other natural disasters, it is naturally the human dimension that is the most tragic. The incomprehensible losses of human life in recent years stand like monuments in the mind to the inability of our societies to protect themselves from danger. Just think of the 280,000 people who lost their lives and the several million who were made homeless in Indonesia, India and Sri Lanka in December 2004. Or the poor and those with few resources who remained in New Orleans in September 2005 – and the fate of all those who managed to get through the earthquake in Kashmir in 2005 with their lives intact but who did not receive enough help to survive the harsh winter that soon followed.

Do natural disasters have an effect on cultures and societies? Can they influence the conceptual world and mentality of human beings? This book presents a number of answers to these questions. During the last decades, natural disasters have become a more usual part of the news in both rich and poor countries: the earthquake in Bam in 2003, the tsunami disaster in Southeast Asia in 2004, Hurricane Katrina in the USA in 2005, the cyclone in Burma in 2008, the earthquake in China in 2008, floods and heat waves in Europe during recent years, plus hundreds of other large- and small-scale disasters round the world. Despite the great attention paid to them, we are rarely given any in-depth analyses of what these disasters do to us. My motivation to write this book has been a belief in being able to make use of what we have learnt from previous disasters so as better to understand the consequences of the natural disasters of our time. When I began to research for this book, my attitude was that natural disasters are mainly caused by forces of nature. In the course of writing it, I started to understand that this is a serious oversimplification at best.

My tales of natural disasters cover wide expanses of time and space. Our journey begins in pre-Christian Europe, then continues towards the present time via a series of disasters in various cultures. It stops at one point where Europe is about to take the leap into the Industrial Revolution – a watershed between the old and the modern where natural disasters were of great importance. After examining disasters in Asia, usa and Latin America, it finally arrives at the great events of our own age – and what we can expect as a result of global warming.

While the book covers wide expanses of time and space, it also takes place at another level, exploring the fate – and the explanations of possible causes – of those who, throughout history, have been the victims of natural disasters. It is a journey that takes place on the borderline between 'faith' and science.

Mythologies

'The heathen gods had no moral scruples.'
Augustine

There were bad years with crops that failed, and the Svea people in Sweden were starving. They hoped that the sacrifice of oxen to the goddess of fertility, Freya, would help the next year's crops and that the granaries would be filled. But the following autumn the situation was just as hopeless. In order to show how far they were prepared to go to regain the trust of the goddess, the Sveas, with King Domalde at their head, sacrificed men. Despite this, the crops still failed and the famine continued. At the time of the third harvest they gathered once more in great numbers in Uppsala to make new sacrifices. Even stronger means had to be adopted if the goddess was to be satisfied, so the chieftains met to discuss what was to be done. The order was given: King Domalde had to be sacrificed. He was hanged, then stabbed, and his blood smeared over the sacrificial altars.

King Domalde was the first king of the oldest ruling dynasty in Scandinavia, the 'Ynglings', in the Viking Age, over 1,000 years ago. The story comes from Snorre Sturlasson's *Heimskringla*, a collection of king sagas, but even though it is mythical, similar sacrifices are thought to have been common throughout Northern Europe during the pre-Christian era. Sacrifices to the gods were carried out several times a year in connection with a number of rituals. The Swedish religious historian Britt-Mari Näsström has pointed out that sacrifices of animals and humans were made in order to avert critical situations and that, as a last resort, the most important persons in society had to pay with their lives.[1] The Norse sagas provide us with pieces of the jigsaw puzzle regarding pre-Christian explanations of – and attitudes towards – natural disasters.

Ragnarok is the tale in Norse mythology of how the gods and the giants, Jötnar, meet in a final battle that ends with the world being destroyed and the gods perishing. Mythologies explain natural phenomena and relate what place disasters occupy in people's conceptual worlds. The first warning of the impending Ragnarok is social and moral decline – and the cruel *Fimbulwinter*. This winter will last three years where there is no summer and lead to famine and great conflicts. The sun and the moon will disappear and the stars fall. Powerful earthquakes will let loose the wolf Fenrir while the Midgard Serpent will lash and whip up the sea to a storm. Huge waves will roll in over the land and tear the ship *Naglfar*, made from the nails of dead men, from its moorings, while the giant Surt battles against the gods, unleashing volcanoes and eternal fire. The destruction of the world draws near. The gods fight bravely, but Surt's volcanic eruptions prove too strong for them.

The battle between the gods and the giants at Vigridvollen mobilizes all of nature's forces and completes Ragnarok. All gods, giants and humans die before the world sinks into the sea. This sinking of the world is described in the poem 'Völuspá', which is more than a thousand years old:

> The sun darkens,
> earth in ocean sinks,
> fall from heaven
> the bright stars,
> fire's breath assails
> the all-nourishing tree,
> towering fire plays
> against heaven itself. [2]

After this, the world resurfaces, re-emerging with both survivors and newly-risen gods. In the new world there will be no volcanoes. Humans and gods can peacefully co-exist without the interference of the giants.[3]

The tale of Ragnarok shows that earthquakes, volcanic eruptions and harsh winters are associated with the end of the world.

Surt controlled the volcanoes, and earthquakes were caused by Lóki twisting in pain where he lay bound, with adder's venom dripping onto his head. Mythologies do not tell us much about people's reactions and attitudes during natural disasters, but it is clear that the fear of natural hazards was great.

BEGINNINGS

Humans have always cultivated and worked with nature, but there is a point where technology and knowledge are no longer sufficient to protect them against natural hazards. What could they do when their measures were not sufficient? They must try to pacify the gods, who, in their worldview, were the only ones capable of reducing the dangers they could not handle themselves. If one goes far back enough in time, practically all cultures the world over explained natural disasters as the punishment of the gods. In the Nordic countries, Norse gods were worshipped in order to reduce natural hazards, while in other parts of the world other gods and rituals played the same part. There are many similarities between accounts of the creation and ending of the world between the Norse, Indian, Iranian and Greek mythologies. Greek and Roman mythologies, for example, had elements of divine punishment in connection with volcanic eruptions and earthquakes. Mythological stories of floods are also common in ancient cultures all over the world, not least in the Old Testament.

The sacrifice of King Domalde is not unique. Other stories from pre-Christian era Sweden and northern Europe show that human sacrifice was a means of re-establishing the trust of 'the irate gods'. One example is the burial of live children. One or two poor children could be enticed into a pit with a little food and then be buried alive. In the *Reykdœla* saga, it states that the sacrifice of children or old people was discussed during a famine, although in the end this was not carried out.

The knowledge we have is often of a general nature, there being little information about reactions to specific disasters. The story of Domalde says a great deal about people's view of natural forces in the pre-Christian era. Most researchers who have dealt with the theme have primarily talked about rituals in connection with famine. These rituals were regular features of the passing year – as, for example, when Domalde was sacrificed at harvest time. We can also assume that sacrifices to prevent floods were among those to the goddess of fertility. But we must expect different types of reactions when it comes to sudden natural disasters, such as landslides, earthquakes or tsunamis. At a general level, we can assume that the disasters created fear, and that people felt in need of an explanation as to why such a thing had happened. Possibly, refuge was sought close to holy places or shrines. If no obvious explanations were forthcoming, the events were interpreted as the punishment of the gods. Through religion, people acquired answers as to why farms had been destroyed or family members had lost their lives. Perhaps people had their own kinds of disaster ritual that could be used, independently of the fixed rituals in connection with sacrifices to Freya or Yule.

There were also magic rituals for counteracting storms and influencing the climate. During long periods of drought, women in German villages would go round the town inspecting their young maidens. One was then selected to walk naked in front of a procession out of the town. There, the maiden would dig up the root of a particular herb, which she then placed on her left foot. The maiden would then have to walk backwards to the village in more or less crablike fashion.

PRE-CHRISTIAN GHOSTS

The transition from pre-Christian to Christian culture was gradual, the popular culture of the medieval period retaining a number of pagan characteristics. Even the official Christian culture had strong traces of pagan rituals and customs.[4] The most important difference was that the responsibility for controlling nature was transferred from the heathen gods to God, Jesus, the Virgin Mary and the saints. The role of the Virgin Mary was to protect people and appease God, while that of the saints was to control the natural forces. In addition, new rituals were created in the meeting of the old and the new. One example of this is the sacrifices made to the god Robigus in connection with Robigalia Day, which had been a pagan Roman ritual to ensure a good harvest, but which was continued as a Christian ritual in parts of Europe. The ceremonies on Robigalia Day could easily be converted into disaster ceremonies if necessary. These processions were intended to be annual preventive measures and could be directed against drought and floods. Robigalia Day is still celebrated by the Catholic church on 25 April each year, in connection with St Mark's Day.

Traditional tales and folklore provide interesting information about how mythological tales were passed on during Christianity. We also gain insight into how people have interpreted natural phenomena and natural disasters, and the tales can have had an educative role for new generations. Through these tales the young hear about the risk of natural disasters in the local environment.

'A ledge of rock lay on the mountain side above the houses on Aamland. If it were to crash down, it would cause great damage to the farm. Every Christmas Eve they offered a bucket of butter to the rock, so that it would not fall down and ruin the farm.' The story is one of many tales recorded in Norway since the eighteenth century, and in this instance shows an offering being made in the hope of preventing a rockslide. Similar stories about popular belief are to be found in all of western Norway. There was a widespread belief that steep mountain sides could collapse and bring about the end of the world. This shows that the fear of rockslides was definitely present, even though the explanations differed from those we have today. It was often believed that people were punished for their sins by natural disasters. An example of something that could unleash a natural disaster was a sinful wedding procession that passed under a cliff. 'When brother weds sister, it will collapse on them. Then the last days have come.'[5] Incest was one of the signs of moral decline that was believed in pre-Christian times to introduce Ragnarok.

After the conversion of Europe to Christianity, the cross was actively used to hold evil spirits at bay. Farmers would place crosses out in the fields to protect the crops, or they could be placed on mountains to prevent storms and bad weather.[6] This custom has survived to the present day.

Folklorists have pointed out that it was common in Norway for the natural forces to be personified as Jutuler, trolls that lived in the mountains – a survival of the mythological Jötnar. The thundering and crashing that often occurs in mountainous regions with large changes of temperature were interpreted as the coughing of these trolls. Researchers do not agree, however, on how the popular belief is to be understood. The Norwegian cultural historian Arne Bugge Amundsen claims that popular belief and Christianity were two sides of the same coin. Popular belief was thus not exclusively a local phenomenon but may have represented widespread attitudes that were an integral part of Christian belief.[7]

MEDIEVAL RITUALS

Present-day attitudes towards natural disasters vary according to country and religion – and, on a smaller scale, according to local natural conditions and the social status of the people affected. If we go back in time, the amount of information becomes less and it seems to become easier to gain an overall view of the handling of disasters. At the same time, it becomes more difficult to know details about how people acted and thought, and how representative the information we have actually is. In the pre-Christian era, we can definitely speak of an 'information underload'. If, however, we move on to the Middle Ages, the amount of source material becomes greater. We know quite a bit about the concrete actions performed in medieval religious rituals designed for natural disasters.

This source material is so good that we can form a picture of how natural disasters have been conceived in Europe since the thirteenth century. It is in particular the pioneering work of the Finnish historian Jussi Hanska that has shed new light on how Christians regarded natural disasters in the Middle Ages. And we learn valuable things about societal organization, mentalities and religiosity into the bargain.[8]

It can be assumed that every generation during the Middle Ages had to consider how one was to survive after a major natural disaster. The concept of 'natural disaster' did not, by the way, appear until well into the eighteenth century as a term for all the problems that had befallen man as a result of original sin. Before then, the concept *tribulatio* had been used. *Tribulatio* can be translated from the Latin as 'unusual adversity' or 'difficulty', and the term included natural disasters as well as other mishaps, and manmade disasters such as wars. The knowledge we have of reactions to natural disasters in the medieval period mainly comes from chronicles, canonizations and the writing down of church services. But the most important source is the so-called disaster ceremonies.

Disaster ceremonies were held in connection with specific natural disasters and were written down so that they could be used in similar circumstances elsewhere. They show what the Church told the common people about disasters. The ceremonies – 'disaster liturgy' – represent formerly unused material for understanding reactions to natural disasters, and include processions and masses.

Christian processions designed to prevent natural disasters can be traced back to the end of the fourth century. But the best-known of the early Christian processions took place in the year 589 AD. It was organized by Pope Pelagius II after a great flood in Rome in order to prevent an outbreak of the plague. The procession was to end in the worst conceivable way, with 70 people dying of the plague during the actual procession. This did not, however, prevent similar processions being organized as long as the plague raged. It later became a model procession for other countries under the name *Litania romana*.

The processions were ceremonial ones, with the crowd moving from one church to another. The idea was for the common people to take an active part in the proceedings, normally walking barefoot while praying. Prior to the procession, alms were distributed to the poor, and it was also usual for holy relics to be carried along during the march. The order was fixed: the clergy in front of the laity, with and married women bringing up the rear. At the final station a mass was organized.

The processions were planned well in advance, as droughts and floods often last for some time. The largest procession we know of was organized in Paris in 1412. Tens of thousands of people took part. During the masses after the processions the priests gave practical advice about the attitude one should adapt to disasters – and, naturally, their view of the causes and how one could obtain forgiveness of sins.

In the event of sudden natural disasters, such as earthquakes, the normal procedure was to seek shelter in churches, protected by what is sacred. If one did not make it to the church, one could perhaps read the *Pater Noster* or *Ave Maria*. Once the disaster was a fact, one could turn to the Bible to try and find some way of understanding things. If one was struck down by a flood, one could read about Noah and the Flood, and if it was an earthquake, the tale of Sodom and Gomorrah could put things into perspective. Many people made an active effort to get events to stop. During a flood in Florence in Italy in 1333, people were so terrified that found it necessary to beat on saucepans to imitate church bells while praying to God for mercy. Church bells were said to have the power to resist evil spirits.

A QUESTION OF GUILT

It was agreed that this flood was the worst disaster in the history
of Florence. Afterwards, the men of learning, theologians,
natural philosophers and astrologers were asked if the flood
had a natural cause or resulted from God's wrath.

This statement was made by Giovanni Villani after the flood of 1333. It shows the typical mixture of religious and natural explanations for natural disasters in the Middle Ages. Villani himself was interested in astrology, but he concluded that God could have prevented the flood if he had wished to, since, after all, he was in control of nature. Villani died in 1348, during the Black Death. Even though naturalist explanations of natural disasters were known in the Middle Ages, the main cause was usually assumed to lie in the relationship between God and mankind.

Naturalist explanations were normally based on Aristotle's theories and often on some by Thomas Aquinas. This particularly applied to earthquakes. They could be activated if steam found in underground cavities could not escape in a natural way, resulting in pressure building up. The outbreaks were at times so violent that they created earthquakes. Later, we will see that this explanation lived on down through the centuries, being just as popular in the eighteenth century. During a sermon in the town of Viterbo north of Rome, Cardinal Eudes de Châteauroux popularized the pressure model to explain an earthquake that took place in 1269. When one heats chestnuts, they will finally split as a result ovthe pressure that has built up inside them. That is what happens with earthquakes, Eudes explained to the congregation. But Eudes concluded that the real cause was God's wrath at the sins of mankind. Looking for natural explanations seemed purposeless. What actions were perceived as being so serious that they could provoke God's wrath? According to the Franciscan monk Barnardino da Busti in the fifteenth century, the worst sins were pride, robbery and sodomy.

Did the medieval way of dealing with disasters help or hinder the reduction of traumas in people? Here, researchers disagree, but in a way the entire disaster ceremony in the Middle Ages sought to make the disaster comprehensible and to obtain forgiveness from God. As such, it may have had a healing effect.[9] On the other hand, though, the Church gave the people and their sins the blame for the disasters – something that was hard for most people to bear, seen in the light of what we now know about psychology and reactions to critical situations. During a procession against drought in Salisbury in England it was said, for example, that 'it is not the land which has sinned, not the mountain valleys; it is we who have sinned; spare us, Lord, grant us rain.' Today, emphasis is placed in psychological post-disaster therapy on *not* giving individuals the blame for what has occurred.

DEATH WEARS BLACK

The Black Death swept across Europe like a horde of raging wild beasts, consuming everything in its path. Everywhere, the unsuspected disease with the horrible symptoms gave rise to fear. People tried to isolate themselves and avoid contact with others. But it was to little avail. When the infection left humanity alone some years later, about 50 million people, or around 60 per cent of the population, had lost their lives.[10] The Black Death was to leave its mark on Europe for hundreds of years and is one of the worst disasters ever to have hit Europe. Bubonic plague probably originally came from India or China, reaching Italy early in the year 1348. The Black Death was to lead to an upsurge of religious processions in order to prevent what in all strata of society was seen to be God's punishment. There are examples of young women sacrificing themselves in line with pre-Christian practice so that the local community might be spared the plague, although incidents of this type were quite rare.[11] But was the Black Death a natural disaster?

The topic 'what is a natural disaster?' and why such disasters occur has developed into an academic discipline where there have been few simple answers. The biological disasters are often long-lasting, and include famines, epidemics and insect infestations.[12] Labelling plague as a natural disaster does not necessarily mean that such events are independent of human involvement. For those affected – and thereby the possible social reactions to disasters – classifications do not mean all that much. For people in the Middle Ages, the Black Death was among the *tribulatio* that affected everyday life in a dramatic and negative way – and it was beyond human control.

There was a widespread belief in the Middle Ages that the world was coming to an end, and that either the Antichrist or Christ would come. At the turn of the millennium and especially around the year 1260, people went around waiting – and waiting. The plague and wars were among the signs that supposedly indicated the coming of the Day of Judgment. One of the movements that emerged from this cast of thought was that of the flagellants, people who whipped themselves. A fourteenth-century monk described self-flagellation:

> [He] took his scourge with the sharp spikes, and beat himself on the body and on the arms and on the legs, till the blood poured off him as from a man who has been cupped. One of the spikes on the scourge was bent crooked, like a hook, and whatever flesh it caught it tore off. He beat himself so hard that the scourge broke into three bits and the parts flew against the wall. He stood there bleeding and gazed at himself. It was such a wretched sight that he was reminded in many ways of the appearance of the beloved Christ, when he was fearfully beaten. Out of pity for himself he began to weep bitterly. And he knelt down, naked and covered in blood, in the frosty air, and prayed to God to wipe out his sins before his gentle eyes.[13]

The movement developed into a strange sect of peasants, craftsmen and vagabonds with a strong belief in the Last Day whose aim was to crush the Church and kill all priests. They challenged the authority of the pope, claiming they were directly led by the

Holy Spirit. When the Day of Judgment failed to come in 1260, the movement gradually died out. After a powerful earthquake in southern Europe in 1348, it reappeared. This earthquake was interpreted by many people as 'messianic labour' that would lead to 'the last days'. The Jews were blamed for the 1348 earthquake, and many were caught and burned. In addition, the Black Death appeared on the southern European scene that same year. The frightful death that bubonic plague led to was considered to be a punishment from God.

With the earthquakes and the Black Death, the flagellants were at the centre of the events they had been waiting for. They claimed that the cure for the plague was processions and the killing of scapegoats. It was first assumed that the cause of the deaths was that the Jews had poisoned the drinking water. They were persecuted and killed. The flagellants collected large crowds of people who went round the streets while scourging themselves until the blood ran. The local population joined in the hunt for Jews. They were virtually wiped out in the large cities of Germany and the Netherlands. A contemporary account soberly describes the horrible state of affairs:

> Plague ruled the common people and overthrew many,
> The Earth quaked. The people of the Jews is burnt,
> A strange multitude of half-naked men beat themselves[14]

No one in Europe was able to live through the 1340s without thinking of the bubonic plague and God's punishment. An analysis of reaction patterns in France during the plague has shown that they can be divided into three. Firstly, philosophers and those versed in natural science believed that comets and the unfavourable constellations of the planets had caused the plague. Comets were always seen as warnings of disasters. The second group comprised ordinary people who were convinced that mankind was responsible for the plague, so they looked for those who were guilty. In many places, the Jews were blamed. If a scapegoat was found, this could restore the balance disturbed by the disaster. Scapegoats could also be Moslems, dissenters, or, a few centuries later, witches. In seventeenth-century Norway, a number of women were burned as witches because it was assumed that they had influenced the forces of nature and caused storms and shipwrecks.[15] It can also be assumed that the hunt for scapegoats was partially motivated by the wish to get rid of political opponents, and so propaganda could also play a role. The third type of response to the Black Death came from preachers who believed that the plague was God's punishment for mankind's sins. Most people were familiar with all three explanations, but preferred to choose the one they believed in most.[16]

For those who lived in the fourteenth century, the Black Death was not necessarily a unique tribulation. There were a number of natural disasters that were less spectacular in size and more local than the Black Death, but that nevertheless made a profound impression on people. For people in the medieval period 'history' was a narrow concept that was often restricted to their own lifetime. There were therefore many other disasters that had just as great an effect on the local community as the Black Death.

But the Black Death had enormous consequences. This disaster may possibly have been one of the greatest turning-points in Western civilization. In addition to the religious reactions and the persecution of the Jews in Western Europe, the plague had major consequences on the division of society. Those who had been wealthy prior to the disaster, who owned a great deal of land and depended on a large amount of labour, found it extremely difficult to get hold of enough workers. There was a dearth of people to cultivate the land as well as of craftsmen. For those who had little to start with, and who survived, the plague was a chance to acquire larger areas of land and to climb the social ladder. An important consequence of this was that Europe entered into a period that ended with the transition from a feudal to a capitalist economy. This time of unrest and upheaval also changed Western culture. The changes can be traced in both art and literature, which began to have a stronger interest in death, decay and sin – but also in sexual pleasures.[17]

Plagues and virus illnesses have ravaged the world repeatedly since the Black Death. In the late 1470s, 15 per cent of the population of England, the Netherlands and France died during an epidemic. Even today, thousands of cases of bubonic plague are reported annually by the World Health Organization.

During the medieval period, disaster ceremonies were common throughout Europe – and they never really died out. After the Reformation in the sixteenth century, many of the old rituals were maintained, even though the Protestants altered their form to a certain extent. The main tendency after the Reformation was continuity when it came to the old customs. God's role in natural disasters was not fundamentally changed by the Protestants. Furthermore, a number of natural phenomena were interpreted as signs or warnings from God, and people believed that it was possible through nature to gain knowledge of the hidden God. In both the sixteenth and seventeenth century, signs were sought in nature – in the air, on land and at sea. Calamities were synonymous with divine punishment. Comets, clouds, deformed children, strange fish, illness, earthquakes and natural disasters were all incorporated into a system where everything cohered – and had a meaning. Historians of ideas have pointed out that a number of these so-called omens were thought to warn of the coming of the Day of Judgment. When the famous Danish astronomer Tycho Brahe discovered a new star in 1572, he claimed that it would bring great misfortune: plague, war and political upheavals would come. A number of countries would be affected, Brahe stated.[18]

From the Middle Ages on through the Renaissance there was an increased interest in the philosophers of antiquity, art and science. In the course of the sixteenth and seventeenth centuries, people gradually began to make their own observations of nature – something that led to a proliferation of new theories. However, Aristotle was still the leading authority when it came to earthquakes, for example, and God was the supreme Mover when it came to natural disasters. Natural historians tried hard to find evidence that the Flood was a historical event, but this new evidence proved difficult to reconcile with the old authorities and also, as the historian Norman Cohn has pointed out in his book *Noah's Flood*, the observations did not always tally with the 'answer book', the Bible. Another important development was that it was seen as less possible that God would interfere in his own creation. The germ of the ideas that were to flower in the eighteenth century was sown by the philosopher René Descartes, who claimed that

God had created the universe, but that after its creation the world was solely controlled by physical laws. With specific laws for the workings of nature, there was less room left for miracles – which do not, of course, obey the usual laws of causation. However, the medieval view of natural disasters continued to survive in nearly all strata of the population and, by and large, it remained unaltered throughout the seventeenth century.[19]

In the eighteenth century greater emphasis was gradually placed on the superiority of the intellect, rationality and science at the expense of the Bible and old secular and religious authorities. The Enlightenment is the name historians have given this period, based on the main changes that took place within politics, philosophy, science and human mentality. Even the old pre-Christian rituals gradually lost their hold. The new way of reacting to the world meant at the same time that the conception of natural disasters shifted from being the work of God to being the result of the processes of nature. Another important change also occurred: God, who previously had been interpreted as hard and punitive, was seen as being loving. Philosophers and theologians claimed that the world was created in God's image and that he controlled it according to a loving, rational plan. At the same time, the idea that nature represented wildness and something barbaric was also more widespread than ever before. An important objective in the Western world was to tame nature, to exploit it and subjugate it, with the hope that this could recreate a Garden of Eden here on earth.[20] Armed with rationalism, new land areas were conquered and 'civilized'.

Powerful earthquakes in Italy (1693), South America (1746, 1783 and 1797) and Portugal (1755) had a drastic impact on this idea of a protective, benevolent God and helped to push forward new explanations of natural phenomena. They were also a shot across the bows as regards the taming of nature. 150,000 people are thought to have perished in these disasters. Debates on God's role in natural disasters raged throughout the Western world. The Lisbon disaster of 1755 is the one that has been best studied, and via the reactions to this disaster we have gained insight into the system of values, beliefs, politics, crisis-handling and view of mankind that prevailed at the time. Before turning to the earthquake in Lisbon in 1755, however, we will take a look at what, in terms of size, was an insignificant earthquake, which took place in England in 1750 – one that nevertheless gave rise to strong reactions.

'Of all the judgments which the righteous God inflicts on sinners here, the most dreadful and destructive is an earthquake.' This assertion comes from theologian John Wesley, the founder of the Methodist Church. Partly underlying Wesley's view may well have been the two small earthquakes of February and March 1750 that created panic in London. Even though the quakes did not result in any deaths or damage to buildings, people were scared out of their wits at the thought that London was unsafe.

People in London were terrified after the February quake. A prophecy that a final annihilating earthquake would destroy the city on 4 or 5 April did not make things any better. Panic broke out, and as many as 100,000 Londoners fled the city for the country. The earthquake did not occur, but many people chose to stay away for a while to be on the safe side.

The Bishop of London, Thomas Sherlock, published a pamphlet in which he rejected the scientific explanations of the earthquake. It was impossible, he claimed, to understand earthquakes without including God and the sins of mankind. The pamphlet sold in large numbers and gave people what they wanted – an explanation. The contributions to the ensuing debate came from clergymen and researchers, covering the entire spectrum from God and the role of creation to an emphasis on natural laws. The reactions have been described by the historian Thomas D. Kendrick, whose book *The Lisbon Earthquake* (1957) is still one of the best studies of the view of natural disasters during the Enlightenment.[21] The clergy had good reason to bring God into the debate. It corresponded to the overall view of the population and, furthermore, created a front against the godless philosophers, Kendrick claims.

The reason for the groundless panic was a belief that everything in the world was for the best. A disaster or an earthquake was therefore a powerful attack on this attitude, indicating that God had a reason for punishing the population. The earthquakes in London led to theological scrutiny, penances and speculations on man's relationship to God. If the reactions after an earthquake that resulted in neither loss of life nor major material damage could be so strong, what might not happen after a true catastrophe? The answer came five years later, when Portugal was hit by one of the most powerful European earthquakes we know (see next chapter).

Even among natural scientists, biblical explanations of the processes of nature were upheld even up to the mid-nineteenth century. (And, as a curiosity: not until the end of the nineteenth century was the supplement removed from Scandinavian and English bibles in which it was stated that the earth was 6,000 years old.) A turning point as regards the view of natural disasters came during a cholera epidemic in England in 1832. An official prayer day with fasting was organized on 21 March in order to diminish God's anger – and the churches were full. But others had completely different explanations. In some towns workers rebelled in the streets because they believed that the deaths were due to the authorities trying to solve the problem of poverty by poisoning. The rumours came from radical working-class newspapers.

ON COMETS AND BACON

The fear of comets, so widespread during the Middle Ages, lived on as the centuries came and went. When Halley's comet approached the earth in 1910, the fear of poisoning was great. It was believed that the tail of the comet, which would enter the orbit of the earth, contained cyanide. While many people bought telescopes to be able to follow the course of the comet through the universe, others were panic-stricken at what might occur. In big cities around the world, such as San Francisco, London, Prague and Tokyo, 'anti-comet' pills sold like hot cakes. On 18 May 1910, the day before the comet was at its closest, The New York Times wrote on its front page that 'if this is the last edition of THE TIMES we wish you a fond farewell.'[22] Many churches stayed open that night. Some people believed that prayer did not offer enough protection against the dangers of the comet. The religious group The Sacred Followers in Oklahoma were arrested by the police when they tried to sacrifice a virgin.[23] The businessman A. Murray Turnerin, of the town of Hammond, USA, was more down to earth. He explained to the local newspaper that he wasn't worried: 'my new house is both bomb-proof and comet proof.'[24] Similar stories, though less extreme, appeared in Scandinavia. The fear of the Day of Judgment was widespread, and many people prepared for the worst:

> I have a vague memory of Halley's Comet in the year 1910, when I was 10 years old. Many people were hysterical and believed in the end of the world. Older people thought that the comet would sweep everything along with it when it passed. We children were naturally frightened and influenced by the fear of the grown-ups . . . We lived next-door to my aunt's family and the day the comet was to pass over, they moved in with us, so that the whole family could be together when the disaster occurred.[25]

In Örebro in Sweden, the preacher Petrus Lewi Pethrus held a revival meeting so that people could convert before the end of the world became a reality. In Norway, a woman who saw smoke some distance north of her home is said to have anxiously exclaimed: 'No, now it's already burning in Kopperud!' Though not everyone took the threat from above equally seriously – a man is said to have stated afterwards: 'If I thought I would perish on Thursday, I wouldn't save on the bacon and my best flour.'[26]

2

The Day of the Dead

'Human beings are not nature's favourites.'
Camille Paglia, *Sexual Personae*

Africa is jostling impatiently at Europe's borders. It is as if the entire continent wants to move northwards. Friction is inevitable and it is impossible to remain unaffected. This has nothing to do with the advance of Islam in the eighth century or with today's stream of immigrants from North Africa across the Straits of Gibraltar to Europe. It has to do with the movement of the continental plates and the geophysical causes of the largest earthquake disaster ever to hit urban Europe. To be more specific, the problem is a collision between the African and European continental plates, and the faults and earthquake zones in the area around the Bay of Cadiz. They affect both Portugal and North Africa, with some of the faults rising as high as the sea bed. This is why earthquakes that originate here can also cause tsunamis. The earthquake and tsunami that occurred on 1 November 1755 was to hit Portugal and northwest Africa hard and to send both physical and intellectual shockwaves out across the Western world.

THE PRESENCE OF DISASTERS

Baixa was the commercial and administrative centre of old Lisbon – it was from here that activities in the Portuguese empire were led. The precinct was almost razed to the ground in the earthquake of 1755. Today, Baixa stands as an impressive monument to reconstruction after the earthquake. The rectangular quarters were erected by Minister Sebastião José de Carvalho e Mello and his architects. Mello was later to be given the title Marquis de Pombal and to become Portugal's most powerful man in the wake of the disaster.

Pombal led the reconstruction of Lisbon, based on a rational and carefully thought out plan. In the new Baixa, the streets were to be wide and the houses dwelt in by craftsmen and businessmen. The various occupations were placed in their own streets. On the ground floor lay the business premises, which today house clothes shops, cafés, antique shops and banks. The upper floors housed dwellings. In Rua dos Bacalhoeiros – fishermen's street – there now lie a string of restaurants, but you will search in vain for headwear in Rua de Sta Justa – the former hatmakers' street.

How many of the tourists who visit Lisbon think of the tragedy that took place here over 250 years ago? The local inhabitants, for their part, have taken care to keep the memory of the event alive. The ruins of the church Igreja do Carmo still stand guard over the city centre as the most prominent icon of the earthquake and the destruction it caused. What mechanisms ensure that the disaster is not forgotten?

Historians of ideas have underlined that what a society can no longer recall can tell us as much as the memories it keeps alive. The Lisbon disaster is one of those that people definitely remember – all over the world. In the period after the tsunami disaster in December 2004, innumerable writers and commentators in both USA and Europe drew parallels with the Lisbon disaster. Despite the fact that the disaster occurred in the eighteenth century, it was still considered to be relevant for understanding how people react to major natural disasters. The Internet can be used in this connection as a kind of yardstick for present-day mentalities. A search on Google in February 2005 produced over 6,000 hits for 'Lisbon 1755'. Interest gradually waned, and in July 2005 this figure had dropped to 600. New events win the competition for our attention and cause even large disasters, slowly but surely, to disappear from focus. Since the disaster in southeast Asia had global consequences, it is probable that it will be remembered for a long time. If the earthquake in San Francisco in 1906 is the natural disaster best remembered from North America, Lisbon in 1755 is the archetype of the European natural disaster. Furthermore, the 1755 event was the first major *modern* natural disaster. The British historian David Birmingham has pointed out that the quake was so important that it is the *only* event in Portugal's history that has imprinted itself on European memory.

Was the Lisbon disaster really so important, or is it always referred to because it is *this* disaster we have chosen to remember at the expense of all others? We ought to keep both options open. An earthquake in 1531 caused just as much damage in Lisbon, though very few people indeed have ever heard of *that* disaster.[1] Reconstruction after the destruction in 1531 was so expensive that the Portuguese had to abandon several of their colonies in North Africa. The number that perished is unknown.

The natural disaster that struck Lisbon in 1755 has been claimed as being among the most dramatic examples of how natural disasters can affect societies, ideas and mentalities. This is probably why the disaster is still known today. The subsequent political changes in Portugal were large and Europe, highly religious at the time, was shaken to its very foundations. Intellectuals threw themselves into lengthy debates as to the causes. It was the question of guilt that was discussed most of all. Could God be behind the natural disaster? And, if so: why did God want to punish the Portuguese? Could it have something to do with Portugal's cold-blooded exploitation of Brazil, or the public executions of the Inquisition?

THE DAY OF THE DEAD (ALL HALLOWS)

The course of events is well known. The most powerful earthquake, measuring 8.4 on the Richter scale, struck Lisbon between 9.30 and 9.50 on the morning of 1 November 1755.[2] The three most powerful quakes caused the city to shake for about 10 minutes. Hundreds of aftershocks shook Lisbon for the following six months. The quakes came from a fault in the sea southwest of Lisbon and could be felt throughout Portugal, Spain and northwest Africa. The main quake caused considerable damage in Lisbon, but it was the Algarve coast farther south that was hardest hit. Faro was totally destroyed, and few towns in this part of Portugal have any buildings older than 1755. Even so, it is the consequences for Lisbon that people know most about. One and a half hours after the main quake, a six-metre-high wave swamped Lisbon. The tsunami not only hit Lisbon but, like the earthquakes, ravaged the entire coast of Portugal and northwest Africa. The wave was 13 metres high in the Algarve and 20 metres around the Bay of Cadiz. As far away as the Caribbean, the aftereffects of the tsunami could be felt. It is not known for certain how many people lost their lives; estimates vary from 60,000 to 100,000. The extent of the damage was carefully measured by the authorities, and we know from the archives that 85 per cent of all the houses in Lisbon were destroyed. This is where the rational account of the disaster ends. The rest is chaos, death, ruin – and a power struggle.[3]

Thousands were buried in the collapsing buildings, while others ran out into the streets. The earthquake caused both small and large buildings in Lisbon to collapse. Contemporary sources say that hysterical masses fought to try and get out of the city, while clambering over crucifixes, images of saints, personal possessions and dead people. The destruction also caused countless fires to spring up and spread around the city. It took five days for all of them to be extinguished.

Something that exacerbated the panic was the time of day the quake took place. The churches were full, since it was All Saints Day. Just think of the panic that arose in London after what was a comparatively insignificant earthquake in 1750. It must be emphasized that it is a myth that panic often breaks out during catastrophes. Panic is an egoistic reaction to a threat of death. It is not until patterns of individual behaviour actually affect other people's chances of gaining safety that one can talk of panic. On the contrary, the most usual pattern in crisis situations is that people try to help each other to get to safety.[4] When the churches and houses in Lisbon collapsed, one must assume that panic really did break out. The survivors streamed from the narrow streets with their ruined wooden and brick houses towards the open squares, the harbour and out of the city. It must have made a strong impression that practically all of the city's 40 churches were destroyed on All Hallows, the day of the dead. People in Lisbon thought the Day of Judgment had come, and many sank to their knees in prayer. Many tried to escape the burning city by sea, only to be met by the tsunami that had moved at high speed through the deep layers of the sea, in towards the coast of Portugal. The harbour itself was crushed and an unknown number of people were killed – either in boats, buildings along the Tagus river or on the quay. An English merchant described the dramatic scene:

I had now a long narrow street to pass, with the houses on each
side four or five stories high, all very old, the greater part tum-
bled down or continually falling and threat'ning the passengers
with inevitable death at every step; numbers of whom lay killed
before me; or which I thought far more deplorable, so bruised
and wounded, that they could not stir to help themselves ... On
a sudden I heard a general outcry. The sea is coming in, we shall

Natural forces shook Lisbon in 1755 – and chaos reigned. The tsunami that rolled in over the Tejo river struck both buildings and boats. The churches were on fire. The disaster has left an indelible mark on European thought.

all be lost. Upon this, turning my eyes towards the river, which in that place is near four miles broad, I could perceive it heaving and swelling in a most unaccountable manner, as no wind was stirring. In an instant there appeared at some small distance a vast body of water, rising as it were, like a mountain, it came on foaming and roaring, and rushed towards the shore with such impetuosity that tho' we all immediately ran for our lives as fast as possible, many were swept away.[5]

News of the disaster was to be spread around the world by couriers who made use of horses and boats. In that way, the disaster became a media event. But in one important aspect this disaster differed from those that took place after the invention of telegraphy – it took several weeks for the events actually to become news. Only after 7–10 days was the destruction of Lisbon known in Spain and the Mediterranean area, and it took almost four weeks for the news to reach Hamburg. On 8 December the news of the disaster had got as far as the British colonies in America.[6] As we will see, the bad news reached Scandinavia on the same day. In that way the accounts of the disaster became retrospective, with a considerable distance between victims and readers, even in neighbouring Spain. In Europe, the first reports exaggerated the extent of the disaster, with many people sceptical as to whether Portugal could really have been so badly hit. These reactions were particularly strong among the English, who at that time were Portugal's most important trading partners and had considerable economic interests in Lisbon.

TENTS AND HUTTED CAMPS

The earthquake was to pave the way for one of Europe's most powerful 'enlightened' despots, the Marquis de Pombal. Portugal's history in the eighteenth century is dominated by the earthquake, with Pombal the axis around which most things turned. He was given the role of minister and he put King José completely in the shade to achieve his aims. But what was Pombal's role in the chaos that followed in the wake of all the destruction? The most important thing, according to Pombal, was 'to bury the dead and feed the survivors'. This statement has remained as a kind of mantra and led to his being remembered as a man of action who wanted the best for the people.

There was much that needed to be done in Lisbon after the disaster. The dead had to be removed from the streets, looting prevented, African pirates stopped and the city rebuilt. Already on the day after the earthquake, Pombal suggested that the bodies should be sunk in deep water off the coast in order to prevent epidemics breaking out. The army erected tents outside the city, and the Portuguese fishing fleet was ordered home with food for the survivors. The authorities assumed responsibility for helping the population. 34 people were executed for looting in the days after the disaster and a search was started for people who had fled from the ruined prisons. But cities are difficult to bring to their knees, and life gradually returned to normal among the ruins. Portuguese warships were sent to the colonies in Brazil, India and Africa to re-establish trade and dissipate the exaggerated rumours concerning the extent of the disaster.

The disaster did not treat all alike. Very few of those who perished were rich. Portugal had become fantastically rich as a result of Brazilian gold, but the people were poor. Many of the houses of the aristocrats were intact, but they still preferred to sleep out in tents or small provisional wooden huts, from fear of new quakes and out of sympathy with King José, who lived out in a tent for nine months with the entire court. The rich repaired their stone villas while the poor moved to the hutted camps that sprang up like mushrooms on the ridges round the city. No less than 9,000 huts were built during the first six months after the quake. But it did not take long before the nobility could live roughly as before, to the annoyance of some: the patriarch found it necessary to forbid wealthy women to wear coloured hats in church instead of the usual black.

POMBAL AND PORTUGAL

Mapping the extent of the disaster began early. Questionnaires were distributed and complete damage reports were drawn up on the basis of the information received. This was to be the first modern example of damage investigation after a natural disaster. After the destruction, however, rebuilding the city was only one of many alternatives for Pombal and his military architects, headed by General de Maia. Relocation and the complete rebuilding of the entire city were also discussed. Finally, the temporary huts were demolished to make way for a completely new precinct, to be built from scratch. The other damaged precincts were repaired.

Work started on the earthquake-protected rebuilding of Lisbon in 1758. The tiled facades of Baixa conceal one of the most important construction. The houses are built of horizontal logs that have cogging joints at the corners. There is another important feature at underground level: the houses are built on piles, using timber imported from northern Europe. These piles have been driven five to six metres into the ground as a foundation to lessen damage from earthquakes. In addition, fire-walls were erected between the houses, and Pombal had soldiers test their stability by synchronized jumping on them before they were approved for use. These features represent the first systematic modern earthquake-protection measures adopted for urban architecture.

What Pombal is most remembered for is the reconstruction and his vision of a new Lisbon. He fought to modernize Portugal, stopped the persecution of the Jews by the Church and wanted to establish a bourgeoisie at the expense of the nobility.[7] He also intervened in the discussion about the causes of the earthquake that was to dominate the social debate during the following years. Pombal gave natural scientists a free rein to investigate the causes of the disaster, which was directly contrary to what the Church and the religious leaders wanted. This was a strategy to put people's minds at rest and prevent them from believing that they had been punished by God. The spiritual was in no way to be allowed to overshadow the practical. On the sidelines stood the deeply conservative and devout Jesuits, disapproving of Pombal's aims. The Jesuits were involved in controlling the Inquisition, which had its headquarters in Baixa. Once a year it organized public executions where unbelievers and political opponents were tortured and burnt at the stake. Protestants throughout Europe condemned this practice as barbaric, and in England the clergy even believed that the earthquake was God's punishment for the Inquisition. The Catholic clergy of Lisbon responded by blaming it on the presence of Protestants. There are stories of Protestants being caught and forced to convert to Catholicism after the earthquake. The conflict between Pombal and the Jesuits was to go so far that he had them expelled from Portugal. He exerted so much pressure on Pope Clemens XIV in Rome that the Jesuit order was dissolved in 1773.[8] Once the powerful and unpopular Jesuits had been got out of the way, Pombal could at his leisure set about reforming the education system and introducing teaching in philosophy and science.

Pombal's role as a hero has since been called into question. It is possible that he managed to exaggerate his own efforts and to rewrite history. Pombal was not alone in dealing with the living and the dead or in creating law and order, even though he gradually took most of the credit. Most of the other heroes were either dead or had been expelled by Pombal. A few of the noble families were brutally tortured and executed. It is even uncertain whether his famous slogan 'bury the dead and feed the survivors' was actually his own, or was coined as part of a political campaign. Even so, Pombal deserved to be credited for having prevented the natural disaster in Lisbon from having become a social and economic disaster. Pombal was expelled from Lisbon after the death of King José in 1777, and died in 1782.

THE BATTLE OF IDEAS

In the period after the disaster, it was not primarily scientific explanations that people were interested in but moralizing, sermons and religious prophecies. An indication of the general level of concern was the selection of a patron saint of earthquakes. The competition was tough, since there were a number of saints who had successfully been used during earlier disasters in Italy, among other countries. The final winner was Francis Borgia, who died in 1572, and had been a Spanish count and high up in the Jesuit order, with strong links to Portugal. Pope Benedict xiv canonized Borgia in May 1756 after an inquiry from King José, mediated by Pombal.

Did having a special saint help? Despite all the aftershocks in 1756, only few of them were powerful enough to cause more destruction. So people were happy with St Francis Borgia, even though many feared the dark prophecies of the preachers in Lisbon. The preachers were to maintain people's fear of repetitions of the disaster. Pombal brought everything down to earth with his attitude that natural disasters are not outside human control and that vulnerability to disasters can be reduced by improving building techniques.

Most Portuguese were deeply religious and accepted the idea of a punitive God. So it was dangerous to assert that the earthquake had been caused by perfectly natural processes, even with Pombal's support. The first attempt to do so in Portugal dates from December 1755 and was penned by José Alvares da Silva. He expressed himself extremely cautiously, underlining that the disaster could possibly have been God's punishment of the Portuguese, but that the earthquake could also be explained by natural causes. One could learn more about the topic through natural science, de Silva suggested, and it was unwise to compare Lisbon with, for example, Babylon in the Bible. In the years that followed, the debate raged in both Spain and Portugal as to the real cause of the earthquake – if God had a hand in it and how one could best protect oneself against future earthquakes. For his part, Pombal did what he could to get rid of the Doomsday prophets. In the autumn of 1756 it was deemed necessary to arrest those preachers who were walking the streets and claiming that God would destroy the city with a new earthquake on 1 November.

The fate of the Jesuit Gabriel Malagrida in particular shocked people at home and abroad almost as much as the earthquake itself – and it says a great deal about how much was at stake in the wake of the disaster in Portugal. Malagrida, known for his energetic speeches and his magical powers, had previously lived for a while as a missionary in Brazil. He returned to Portugal in 1754 to prepare the queen mother for her death. It soon transpired that Malagrida, who was not known for his tact, came into conflict with Pombal and his supporters. After the death of the queen mother in 1754, Malagrida had to depend on cultivating his friendship with Pombal's enemies. Even though Malagrida was disliked by Pombal as the somewhat extreme preacher that he was, the relationship between them worsened dramatically after the former published a pamphlet in autumn 1756 about the 'real' causes of the earthquake. In it he wrote that it was the people's sins and not comets, underground fires or anything else that was the cause of all the death and destruction. It was God's punishment. And the Portuguese ought to be careful, since they spent so much time on practical things instead of praying to God for forgiveness, he claimed. His timing was terrible, since Pombal was then focusing on scientific explanations and planning the reconstruction of the city. The pamphlet was a direct challenge to Pombal, who could not afford to ignore the opinions of the popular and uncompromising priest. The atmosphere in Lisbon was jittery. Pombal's reply was to expel Malagrida from the city.

Malagrida did not give up, continuing his activities from a new location. The only possibility Pombal had of winning the battle of ideas once and for all was to remove Malagrida from the scene and, at the same time, put a stop to similar attempts. After an attempt on the king's life in September 1758, Pombal got his chance. An earlier letter written by Malagrida, in which he forecast that the king would be injured, fell into Pombal's hands. Malagrida was thrown into prison, accused of attempted murder and eventually handed over to the Inquisition. On one day in September, he was dragged out onto the Rossio square in Baixa, where the Inquisition held its executions. After having been tortured for a whole day, the seventy-year-old Malagrida was strangled and then burned before those who were present. At the same time, a number of Pombal's other political opponents were executed. The remains of Malagrida's body were thrown into the sea. But it proved harder to get rid of the pamphlet about the earthquake. It was still in circulation, spreading fear amongst its readers. Ten years after Malagrida's death, the pamphlet was declared blasphemous by the king and forbidden. Pombal finally appeared to gain control over the interpretations of the disaster. But, ironically enough, banning a book always increases its popularity and ensures that the 'dangerous' texts are distributed and read.

THEORY AND PRACTICE

It was not primarily the news of the number of people who lost their lives that shocked Europe in the months after the earthquake. The biggest shock was that the disaster had taken place at all. This was because eighteenth-century Europe was characterized by an optimism and belief in the future that was punctured by the disaster. Both the aristocracy and the common people believed that they were living in a world where everything had a meaning, where everything was for the best. God had created the world and subsequently controlled it in such a way that nothing was left to chance. The foremost spokesman of this point of view was the philosopher Gottfried Wilhelm Leibniz. He was supported by the poet Alexander Pope, who claimed that there was order underlying all the chaos and chance, and that there existed an indisputable truth: 'Whatever is, is right.' Leibniz claimed that God, when creating the world, had determined everything that would ever happen on the earth, including evil, in a just, loving plan. Good would dominate over evil, so that a world that included evil would in fact be better than one without. Leibniz believed that there were worse things to worry about than earthquakes: 'A Caligula, or a Nero, has caused more evil than an earthquake.' It is probably true that such belligerent leaders have caused more death and destruction than natural disasters but, naturally enough, people still questioned Leibniz's philosophy.

The greatest salvos came from philosophers who saw the earthquake as a chance occurrence and wanted to exploit the disaster as a way of getting rid of the 'all is well' doctrine once and for all. Why did the earthquake hit Catholic Lisbon – and, what is more, on a religious festival with it churches full? Could it be that one did not live in a world where everything was for the best? The French philosopher Voltaire made an attack on Leibniz and on optimistic philosophy in a poem he wrote a few weeks after the earthquake. Voltaire was a well-known person in Europe and many people read the poem, which concluded with the assertion that people are weak, without any knowledge of either their origins or their fate. In his opinion, it was obvious that there was evil in the world. It was stupid to believe otherwise, and the only thing to do was to hope that things would get better in the future.

Voltaire was not the only prominent Frenchman with views on the disaster. The philosopher Jean-Jacques Rousseau was one of the earliest to claim that the natural disaster had social causes. He claimed that the causes lay in people having chosen to live in cities and in large houses, instead of in small houses in the country. So the disaster could have been avoided if society had been organized in a different way, Rousseau asserted. An exchange of words between Voltaire and Rousseau has since become a classic for our understanding of how societies and disasters were viewed in the eighteenth century. As a consequence of the earthquake, divine providence was rejected by the intellectuals, while the belief in progress, science and rationality was temporarily checked. Voltaire did not content himself with a poem in order to promote his views. He also wrote an entire novel satirizing optimistic philosophy, *Candide*, which became a bestseller that gave optimism the *coup de grâce*. Both Leibniz and the attitudes of the Inquisition to the earthquake were subjected to criticism:

> After the earthquake, which had destroyed three-fourths of the city of Lisbon, the sages of that country could think of no means more effectual to preserve the kingdom from utter ruin than to entertain the people with an auto-da-fe, it having been decided by the University of Coimbra, that the burning of a few people alive by a slow fire, and with great ceremony, is an infallible preventive of earthquakes . . . Candide, amazed, terrified, confounded, astonished, all bloody, and trembling from head to foot, said to himself, 'If this is the best of all possible worlds, what are the others?'[9]

GOD OR SCIENCE?

What in the way of natural explanations could one use after having rejected divine influence? It was not until after the earthquake in San Francisco in 1906 that a theory we now recognize as sound was launched. In the 1750s the understanding of geological processes was quite different from that of today, and the most common theories explained earthquakes by fires in underground sulphurous veins and the collapse of underground cavities. Scientific theories of this type were in essence based on the philosopher Aristotle's theories from antiquity, which enjoyed a new upsurge of popularity in the sixteenth century.

An interesting view of the natural disaster – one typical of the age – came from the German philosopher Immanuel Kant, who published three articles about the earthquake in 1756. Kant was not particularly interested in the event as a human tragedy but in the scientific aspects. He was not convinced that the earthquake only represented something negative. Aren't we all going to die? Earthquakes are part of nature, and we have to adapt. We cannot expect nature to adapt itself to what is best for us, he claimed. In his opinion, the earthquake could not be related to the moral state of humanity. Why God had allowed the disaster lay outside human comprehension. But Kant's explanation of the actual earthquake was naturalistic and based on his studies of the literature of natural history. Like many others, he believed that flames from the interior of the earth had forced a path up to the outer crust through a labyrinth of passages and channels. Fire was explosive, and was ignited by water meeting sulphurous areas in the earth. This led to earthquakes. Kant placed special emphasis on the tsunami that surged in over Lisbon, explaining it as a propagation of earthquake waves through water. He also made an important personal contribution: the sources of an earthquake can lie a long way from where they are actually experienced through the creation and movement of waves.

The Lisbon quake led to important advances within science. What set things going was the questionnaire scheme started in Portugal after the disaster. The so-called 'Marquis de Pombal survey' contained 13 questions that focused on what people had observed in the way of physical reactions, such as the number of tsunamis, aftershocks, material damage and the number who lost their lives. After the disaster, the British physician John Michell launched a hypothesis on the propagation of tsunamis and earthquakes. The shock waves move more slowly through water, something that can explain the observation that the tsunami to hit Lisbon arrived after the quake itself. Michell was a transitional figure between the ancient and the modern who helped to contribute to the theories of the Greek philosophers being laid aside. Michell's theory on the propagation of waves was sound, even though he explained earthquakes as having arisen from underground fires and explosions – not from the building up and release of tensions along faults.

TREMORS ON THE EDGE OF EVENTS

There were unusual occurrences at several places in Europe after the earthquake. The repercussions of the Lisbon earthquake were felt in Norway some hours after the disaster. The Norwegian coast was hit by a tsunami, even though it lay 3000 km from the epicentre of the quake. A huge wave suddenly rose up from the quiet waters of a fjord in western Norway, and struck the shore and flooded about ten metres inland. In the space of fifteen minutes, it came back twice more.[10] As no one had heard reports yet on the

Lisbon disaster, the events were regarded as isolated incidents that had no explanation. Other observations must be taken with a pinch of salt, such as accounts from Norway of flames and balls of fire careering through inland lakes.[11]

Unlike the physical reactions, it took a long while for the news to arrive in Scandinavia. Not until five weeks later, on Monday 8 December, was *Kiøbenhavnske Danske Post-Tidende* able to inform its readers of the tragic news on its entire front page.[12] 'Via a courier message from Lisbon arrived her today at 4 p.m., we learned that in that same place ... an earthquake had been experienced that had led to the most unfortunate of consequences.' Most of the article dealt with how the royal family in Portugal had tackled the difficult situation. The newspaper wrote that the royal family had had to manage for a whole day without servants and that they had had practically nothing to eat. There were also accounts of how other important personages such as counts and ambassadors had fared. The victims of the disaster were given a sentence at the end of the supplementary article in which it stated that '50,000 inhabitants of Lisbon have lost their lives in this calamity'. On the same day, news of the disaster was printed in newspapers in the American colonies. In Boston the information came from a ship's captain by the name of Johnson who himself had been in Lisbon during the disaster.

The west coast of North America was hit by an earthquake on 18 November 1755, so many people were most interested in learning more about the causes of earthquakes. The *Boston Gazette* published an explanation on 24 November: 'Surely no one will question the Agency of the Supreme Power who maketh the Earth to tremble and whose voice shaketh the Wilderness.'[13] That set the discussion going, long before people had even heard of the disaster in Lisbon.

These reactions echoed the debates that were raging in Europe. Intellectuals threw themselves into exchanges of words regarding the causes of the earthquake, and as to whether God, nature or man ought to be blamed for the catastrophe.[14] Old theories of physical causes of earthquakes were recirculated, and a number of new hypotheses were launched. Researchers, businessmen, editors, clergy and kings – everyone wanted to try to understand the disaster.

What could be done for the victims of the Lisbon disaster in Scandinavia? King Fredrik V proclaimed a day of prayer on 14 May 1756, both in Norway and Denmark, to grieve for all those in Portugal who had lost their lives, and, perhaps most importantly, to thank God that the kingdom had been spared and to prevent new disasters.[15] To be on the safe side, a day of prayer was also organized on Iceland on 22 October 1756. One of the central natural historians in Denmark-Norway at the time, Bishop Erik Pontoppidan, made his own contribution with a comprehensive analysis of the earthquake.[16] Could he manage to avoid theological explanations?

Pontoppidan came from a family of theologians, rising to the position of Court Chaplain in Copenhagen at the age of 37. In addition, he was a professor of theology and was jointly in charge of the system of poor relief in Copenhagen. Pontoppidan was to become one of the most important representatives of state pietistic Christianity in Scandinavia in the eighteenth century. As a clergyman, he served at the court of King Christian VI, who was a great supporter of pietistic Christianity. After his father's death in 1746 Frederik V became king. He sent Pontoppidan away from Denmark in an attempted to tone down this same state pietism. Pontoppidan accepted an appointment as Bishop of Bergen.[17] The seven years he was to spend in Bergen were to be highly productive – years when he cultivated his great interest in nature and natural history. He systematized his observations, publishing in 1752–3 one of the first overviews of Norway's natural history.[18] This was the heyday of the impressive attempts to classify all the components of nature, headed by the great masters Buffon and Linnaeus. Despite the fact that Pontoppidan was unable to match the leading researchers of his age, he was very much up to date on most of what was taking place in contemporary research, corresponding with several of the best-known researchers in Europe. Pontoppidan had to leave Bergen for Copenhagen in 1754 to defend himself against rumours that he had made the mayor's daughter pregnant. There was no proof forthcoming, and he was made vice-chancellor of the university in 1755. He never returned to Bergen or Norway.

In Copenhagen, Pontoppidan heard the news of the disaster in Lisbon and began to write a dissertation that he completed roughly half a year later, in June 1756. The book was in the form of a letter to a distinguished, anonymous 'Madam'. He wanted to help alleviate the fear she was experiencing as a result of the disaster. What does Pontoppidan say about the Lisbon disaster? He states early on that on the basis of *morality* he would not have guessed that God would lay Lisbon waste because of its inhabitants' sins. There were other European cities that had deserved the disaster more than Lisbon, without him knowing anything about 'what kind of rod of correction he [God] is preparing for their chastisement'. An important point is that Pontoppidan questions whether the earthquake in Lisbon is an isolated incident. For it would seem that in the years leading up to 1755 there had been an increase in such natural phenomena as earthquakes, hurricanes, 'sulphurous steam', 'boiling waters at certain locations in the sea', and thunderous noises in the ground. Pontoppidan presents a list of natural disasters and phenomena from 1750 to 1755. This increase in disasters had led to anxiety in a number of countries, he continued, and people everywhere were looking for good explanations. It was as if the order of nature had been disturbed.

As for many other theologians and clergymen in the eighteenth century, there was no contradiction for Pontoppidan between a literal interpretation of the bible and a study of nature. Rather, the converse was true – he viewed it as his duty as a clergyman to study nature in order to gain a better insight into God's creation and works. Pontoppidan argued that one had to study how nature behaved in order to understand what God really desires. The increase in natural disasters and earthquakes has to be interpreted as God's will, since the Lord was the prime mover. Pontoppidan's pessimistic conclusion was that one had to be prepared for even more volcanic eruptions and earthquakes. The moral decline of the age and widespread criticism of religion were combined with natural hazards in a spiral of events that would end with the total annihilation of the earth. How long one would have to wait for the Day of Judgment, Pontoppidan did not know. What, then, was the solution? Well, the only thing one could do was to pray.[19]

Pontoppidan always used physical principles to explain the workings of nature, even though God was the prime mover and the first cause of such events as earthquakes. Earthquakes were explained by the collapse of underground cavities and fires in the earth's interior. If pressure could not be released in the form of volcanic eruptions, the crust of the earth would finally split and crack, resulting in an earthquake. Seen as such, volcanoes were positive. They were the earth's safety valves. In addition to the physical foundation of Pontoppidan's understanding of nature we find another feature characteristic of the age. He saw the Flood as a historical event and assumed that the development of nature must have taken place in the space of a few thousand years, driven forward by disasters. He believed it to be madness for atheists to believe in an eternal world and not in an eternal God. If the earth had been formed billions of years ago, he argued, all the mountains would have eroded away and the landscape would be completely flat.

With his belief in the Flood and that the earth was only a few thousand years old, Pontoppidan would today have been called a 'creationist'. Creationists claim that the earth and all life on it were created by God out of nothing and that all forms of life are unchanging. Bible-based natural research contrasts starkly with scientific theories about the development and age of the earth and life on it. The term 'Creationist' includes several Christian groups that have varying views on such fundamental concepts as truth, science and scientific proofs.[20] Creationism arose in the mid-nineteenth century as a reaction against religious explanations of nature being forced out of the scientific world picture. Over a century later, such books as *The Genesis Flood* (1961) by the Americans John Whitcomb and Henry Morris and *Scientific Creationism* (1974) by Henry Morris were to be highly influential in kindling renewed interest in Creationism. Since then, Creationism has been widespread, especially in the USA. Creationists belong mainly to evangelical persuasions, particularly the Pentecostals, Jehovah's Witnesses and Seventh Day Adventists. About half of the population of the USA believes that man was created by God about 10,000 years ago.[21] Similar attitudes, surprisingly enough, are also widespread in Europe. A survey conducted in England in 2006 showed that no less than 39% of those asked supported religious explanations of the origins of life.[22] So there is an unbroken line from Pontoppidan and the old natural historians to the Creationists of today. It is the evangelical persuasions that still have the strongest belief in natural disasters being a sign that the world as we know it is coming to an end.

THE FUTURE

The natural disaster of 1755 left its mark on Lisbon, on history, on architecture and on people. If you ask people in the streets of Lisbon today, they are guaranteed to know about the 1755 disaster and probably are worried about new earthquakes. But what can they do? Precious little, they would say. They are unprepared for new disasters. The Portuguese are fatalistic and always believe that God is on their side: 'It'll all turn out right in the end.' What would happen if an earthquake were to send a new tsunami in over the Portuguese coast?

After the so-called Carnation Revolution of 1974, when Portugal's dictator Camilo Caetano was deposed, many people moved out into the Lisbon suburbs to new and more modern apartments. Many of the shabby and dilapidated town houses were taken over by businesses. Only a handful of families stayed on in the precinct. Today, it is an important objective for urban planners to get people to return to Baixa in order to preserve traditional city life. Would Pombal's Lisbon be able to withstand serious new earthquakes? Why was Baixa hit so hard by the earthquake when other parts of the city managed to avoid such destruction? The answer lies in the ground. This part of the city lies in a flat section between two areas of high ground where two small rivers used to run. They now run through pipelines, but the sediment deposits from the rivers are soft and full of water, so seismic waves cause more damage there than in areas of solid rock. Time after time in history it has been demonstrated that if houses are built on solid rock, they do much better in earthquakes.

Researchers continue to analyse the 1755 earthquake, and in recent years the Portuguese have made great efforts to gain new knowledge about the background to the disaster. Risk maps have been made and calculations undertaken in order to find out how much time will elapse before another earthquake that is more than 8 on the Richter scale will hit Portugal. It will probably take several hundred years for the tensions to have built up to such an extent that this will happen. But happen it will. This is why the increase in tourism and construction along the coast is a cause of concern. No less than 10,000 people lost their lives, for example, in an earthquake and tsunami in Agadir in Morocco in 1960. But it is important to see the risk in perspective: Italy, Greece and Turkey all have a greater risk of experiencing an earthquake than Portugal. In the meantime, attempts must be made to reduce the vulnerability of densely populated areas to earthquakes – and this is the most important lesson we have learnt from Pombal.

3

California: Earthquake and Culture

'A draught of wine would make a mountain dance.'
Omar Khayyam, *Ruba'iyat* [1]

The mountains to the east and west are seldom visible through the thick mist that fills Imperial Valley in south California. Agriculture in what was originally desert is maintained with the aid of irrigation systems that end up in the inland lake of Salton Sea. About a century ago, someone had the bright idea of leading water from the Colorado River into this countryside. Salton Sea was born after powerful floods in the early twentieth century exceeded the capacity of the embankments. This resulted in an inland lake with no natural outfall – one that today is highly polluted, yet is still a stopover for a large number of migratory birds.

From its beginnings along the mountains on the east side of Imperial Valley, close to the Mexican border, the San Andreas Fault stretches northwards for more than 1,200 km. The world's best-known fault is notorious for its powerful earthquakes and it has shaped the landscape of California for the past 20 million years. It is one of many faults that ruthlessly slice their way through California. Several of the faults merge here at the inland lake, the area being one of the most geologically active in the world. There are earthquakes here every single day. Normally, there are 50–60 quakes a week, although practically none of them is due to the San Andreas Fault.[2] Most of the quakes here in southern California come from the San Jacinto Fault.

There was a series of volcanic eruptions here some thousands of years ago, with four small volcanoes serving as a warning that the earth's crust is in the process of rupturing. A number of CHP plants and pools of bubbling clay along the Salton Sea bear witness to this volcanic activity, causing the landscape to seem curious and strange. Carbon dioxide is quite literally spewed out of the ground from the hot clay cauldrons.[3] At all levels, it is an area of great contrasts: agriculture and desert, CHP plants and bubbling clay, large-scale farmers and Mexican immigrant workers. In addition, 'Slab City' lies nearby, a collection of 'winter birds' that have left the cold further north with their mobile homes and settled on a vacated army base – not forgetting California State Prison outside Calipatria.

The American historian and writer Mike Davis relates in his book *Ecology of Fear* how the view of disasters and fear has developed in Los Angeles – a fear both of nature and the black lower class. One way of dealing with fear is to build secure prisons. California State Prison opened in 1993 and it houses 1,200 condemned murderers – and is surrounded by a 5,000-volt electric fence. Contact with the fence means instant death. But it was not human rights organizations that protested after the opening: the migratory birds found the fence fatally attractive, so the authorities had to deal with societies for the prevention of cruelty to animals. After CNN had made a trip to the prison, an ornithologist was commissioned to rebuild the fence. This resulted in the world's only bird- and environmentally-friendly death-fence.

Should you ever visit Imperial Valley, make sure to include Salvation Mountain, due east of Niland. Since the 1980s the devoutly religious Leonard Knight has extended one of the steep slopes outside Slab City with clay. The dried clay has been decorated with layer upon layer of paint in the form of interwoven flowers and biblical quotations. According to Knight, Salvation Mountain is 'a tribute to Jesus'. It is said that he has so far used more than 280,000 litres of paint. But best of all: you can place your finger on the San Andreas Fault if you like. It lies like a dip with a small plain along the edge of Imperial Valley. Here, the boundary between the North American Plate and the Pacific Plate lies exposed. The ground is devoid of vegetation, reflecting the light in shades of grey, brown and orange. The area is known as Painted Canyon, and is typical badlands, bordering on desert. In his book *America*, the French philosopher Jean Baudrillard writes lyrically about the desert colours as products of a timeless geological combustion. In Painted Canyon time has accelerated. Processes that break down rock into clay take place at turbo speed, despite the fact that we as onlookers are only able to see the static

and the eternal. An earthquake in Painted Canyon would be one
without tottering buildings and destruction. In other words: nature
in the raw, without any societal interference. It is friction between
the two stiff pieces of the jigsaw puzzle of the earth's surface that
causes both the earthquakes and the volcanic activity along the
west coast of the USA. But some time has passed since the San
Andreas Fault last moved. It is storing up energy for the next big
earthquake.

The desert oasis of Palm Springs, just north of Painted
Canyon, is a playground and recreation area for the wealthy from
Los Angeles; luxury mixed with faults and earthquakes. A quake
that was 5.6 on the Richter scale shook the town on 8 July 1986,
although no people lost their lives. The houses had been built to
withstand earthquakes. The biggest quake to hit the town in recent
times was caused by the San Andreas Fault on 28 June 1992.
However, it is the San Jacinto Fault west of the town that is the
most active. How can people here reduce the risk of living in an
earthquake zone? Southern California Earthquake Center has
issued a brochure telling you what *you* can do. Point 1: Identify
potential hazards in your home and try to fix them. Point 2: Create
a disaster-preparedness plan. Point 3: Prepare disaster-supplies
kits. Point 4: Identify your building's potential weaknesses and
begin to fix them. Point 5: Protect yourself during earthquake
shaking, 'drop, cover, and hold on'. Point 6: After the earthquake,
check for injuries and damage. Point 7: When safe, continue to
follow your disaster-preparedness plan.[4]

In California, a real high risk of natural disasters blends with
fiction. Natural disasters are exploited by a disaster-hungry media
industry so that we can let ourselves be terrified and fascinated by
natural forces. Sober disaster plans mean little when society as we
know it is laid in ruins.

SYMBOLIC SACRIFICES

It is smouldering beneath the manhole covers of Los Angeles. Gas
explosions and earthquakes spread fear. It gets even worse when
bombs of lava are hurled from the ground in the centre of the city.
The two heroes, an emergency official and a geologist, are the only
ones who can save the city from annihilation.

The Day of Judgment? If nature delivers the final blow, humanity is doomed. A meteorite impact is the worst-case scenario – and therefore often a theme in popular science fiction literature.

The fear of a volcanic eruption in Los Angeles is given the full treatment in the film *Volcano* from 1997. Since Los Angeles does not actually have any volcanoes, the film makes use of an imaginary situation where the threat to the metropolis is the central theme. Disaster movies have always made use of the trick of introducing new dangers into environments one recognizes oneself in.[5] Los Angeles can, on the other hand, offer other dangers than volcanoes. The earthquake of 17 January 1994 at Northridge, in the northern part of the city, led to 57 people losing their lives and over 8,000 being injured. The material damage is estimated at having cost $25–40 billion. After the winter storms of January 2005 Los Angeles was hit by floods, and landslide warnings were issued by us Geological Survey.[6] Large-scale fires have also regularly wrought havoc in southern California in recent years.

Volcano is not unique. Los Angeles has been sacrificed on innumerable occasions. Mike Davis has analysed popular culture throughout the twentieth century and shown that the city has been destroyed in the most incredible ways in films and novels.[7] Nuclear weapons have annihilated the city no less than 49 times and earthquakes flattened it 28 times. A little further down the list we find monsters, devils, fog and storms. Davis claims that the ritual sacrificing of Los Angeles is part of an 'evil syndrome' that has been supported by some of the darkest forces in American history.

In August 1996 a spokesman for the Nation of Islam in Chicago, Louis Farrakhan, stated that an earthquake would hit California within 30 days. Allah's wrath would wipe out Los Angeles. The curse was triggered by a conflict between a property owner and the Moslem tenants of the Mohammed mosque in Inglewood in central Los Angeles. The tenants were finally evicted by the police and the FBI. Los Angeles survived the ordeal, but you do not need to be a fortune teller to predict small earthquakes in California. Christian conservatives have reacted against the liberal lifestyle of California for a long time. Natural disasters are seen by some as a sign that God is punishing the state for its residents' infidelity and for their high divorce rate.[8]

SAN FRANCISCO 1906

From Palm Springs the San Andreas Fault stretches westwards north of Los Angeles. It then moves northwards again via San Juan Bautista to San Francisco, where it ruthlessly slices through the western part of the city, curving out into the Pacific at Cape Mendocino. The San Andreas Fault gets its name from an inland lake south of San Francisco that fills a depression in the terrain, formed by the same fault. Along with the Hayward Fault to the east of the city, the deformation zones of the earth's crust form an 'pocket' round San Francisco. To drive northwards from Mexico along Highway 111 and Highway 1 is one long road trip of meetings between nature and man. Earthquakes, floods, droughts, landslides and fires. Nature and culture.

At 5.12am on 18 April 1906 the population of San Francisco was woken by a powerful earthquake. A few seconds later a new quake came that was even stronger. Hotels, restaurants and business

A refugee camp in the aftermath of the disaster in San Francisco in 1906. Was it the earthquake or the fires that caused the greater destruction?

premises collapsed, with large parts of the city centre ending up as ruins. Shortly afterwards, the flames took over, with more than 28,000 houses recorded as destroyed by fire when the fires were eventually extinguished three days later.[9] Out of the ashes came the result of the worst natural disaster ever to hit the urban USA came to light. One of the world's richest cities lay in ruins. It is not known for sure just how many people lost their lives, but the estimate is over 3,000. It was a release of the tension that had accumulated in the northernmost 430 kilometres of the San Andreas Fault that had caused the quake. Fences and roads moved up to six metres at some points.

The earthquake in San Francisco in 1906 is the best-known of all earthquakes in the Western world, alongside the one that hit Lisbon in 1755. Both quakes are normally mentioned in accounts of the most dramatic natural disasters, even though the number of those who perished is less spectacular than many others. No less than 80 books were published in 1906 that dealt with the disaster in San Francisco.[10] Another reason why the disasters are remembered so well is that both lie in areas where there is the danger of a repeat

No quarter given. An imaginative representation from a French newspaper of how thieves were dealt with during the state of emergency in San Francisco. Looting – or a struggle to survive?

Le Petit Parisien

SUPPLÉMENT LITTÉRAIRE ILLUSTRÉ

DIRECTION : 18, rue d'Enghien (10e), PARIS

À SAN-FRANCISCO
LA MILICE FUSILLE LES PILLARDS DANS LES RUES DE LA VILLE

performance. It has been said of the disaster in San Francisco that it has acquired the status of an icon as the most significant earthquake event we know of. History is compressed, and other great disasters are forgotten. For many people, this event is the one that defines the very concept of 'disaster'. Even so, it appears that it did not do much to improve measures taken to ensure buildings against earthquakes in the San Francisco area. Why? An American professor in the history of disasters, Ted Steinberg, claims that the danger of earthquakes was deliberately toned down for economic reasons.

Was it the earthquake or the fires that caused the greater disaster? Even though the business district was razed to the ground, capitalists and leaders of industry blamed the fires. It is likely that no one would be interested in San Francisco if it came out that the area was susceptible to earthquakes. Campaigns were set in motion to present the city as being seismically safe. Both Governor George Pardee and the secretary of the California board of trade, Arthur Briggs, underlined that the city had been destroyed by *fires* and not earthquakes. The newspapers followed suit. The *San Francisco Chronicle* repeatedly reminded its readers that the most important thing was to rebuild the city. The earthquake must not interfere with that goal. At a meeting one week after the catastrophe, the San Francisco Real Estate Board approved the motion that the disaster was to be referred to as 'the great fire' and not 'the great earthquake'.[11]

It must be said that knowledge of earthquakes in California was distinctly inadequate. The last major quake had taken place in 1868 and knowledge of the geology of the area was limited. Governor Pardee quickly set up a scientific commission to investigate the earthquake, but refused to grant money for research. Money had to come from private sources. The result, the so-called Lawson Report, was published in 1908 and contained a wealth of new information about the geology that lay behind the disaster. It was demonstrated, for example, that damage to buildings was related to the geology of the subsoil. The parts of the city that had been hardest hit were those built on loosely consolidated sediments, as in Lisbon in 1755. One of the most important contributions in the report came from the geologist Henry Fielding Reid. He launched a new explanation of the causes of earthquakes, the so-called 'elastic-rebound theory', which still holds good. It is based on the idea that energy in the earth's crust can accumulate through time and be stored as tension. An earthquake occurs when the rocks in the earth's crust cannot tolerate any more tension and snap back to their undeformed shape, in the same way as an elastic band that has been pulled too far will snap back.

After a powerful earthquake in 1933 along the Newport-Inglewood Fault, focus was once more on California as an earthquake-risk state. A number of schools were destroyed and it was finally felt to be imperative to change legislation in order to improve building standards. The magazine *New Republic* felt that the earthquake might have been God's work, but that the number of lost lives could be blamed on the greed of California business-men. The tide had turned, and seismologists and engineers had managed to convince politicians that economic development was not possible without acknowledging the risk of earthquakes. Even so, the *Los Angeles Times* wrote as late as 1934 that 'No place on Earth offers greater security to life and greater freedom from natural disasters than Southern California.'[12] Yet a report by the Fireman's Fund in 1936 revealed that there were commercial interests behind the minimizing of the risk of earthquakes.

POPULAR CULTURE AND EYE-WITNESSES

The 1936 film *San Francisco* marked the arrival of disasters in Holly-wood. Clark Gable (as Blackie Norton) co-starred with Jeanette MacDonald (as Mary Blake) in a musical that culminated in the 1906 earthquake. According to the historian Ted Steinberg, it was no coincidence that the film was shown in 1936, the same year as the publication of the revealing report. The cat was already out of the bag.

In the film, Mary and Blackie are in a nightclub – a low dive – on the night of 18 April, and the faithless Blackie has just been in a scuffle with Father Mullin. Both of them are Mary's friends and admirers. Unfortunately, Father Mullin is unable to declare his love to her because of his calling. The theme is a classic one: Father Mullin fulminates against lowbrow culture and wants to get Mary away from her job at Blackie's nightclub. Her golden voice ought instead to be heard singing opera.

The earthquake comes, taking up more than 20 minutes of the film – using the most outstanding film effects then available. The earthquake scenes were in fact so frightening that even the actors had to leave the stage from time to time during the takes in order to seek shelter.[13] In the chaos, neither Blackie nor Father Mullin know where Mary is – or if she has survived the disaster. But in a park full of refugees Mary has found her place as a Good Samaritan as she stands there singing for those who have died. Blackie wants to thank God when he sees that she is alive, and the film concludes with 'Glory glory hallelujah' sung by everybody. The fires have been put out and we get a glimpse of the San Francisco of the future being built in the background. Could it be that the director had been inspired by stories from the refugee camp in Golden Gate Park?[14]

Back to the post-quake reality of 1906: on a small green hill in Golden Gate Park, a white-haired preacher collected a crowd of listeners. Despite the disaster he had not forgotten his duties, and a cross and a couple of candlesticks provided a kind of platform. The melody streaming from the cornet of a young man at his side spread out across the camp. Hundreds of people left their tents and moved towards the music. They were gathering in order to worship God.

In the silence that followed, the sound of weeping could be heard. The white-haired preacher continued with closed eyes and raised hands: 'Let us pray.' The congregation fell on its knees before him. A commentator wrote that no one who was in Golden Gate Park on that day could remain unaffected by the prayers.

Charles B. Sedgwick went round the centre of San Francisco, writing down what he saw on the day after the disaster in 1906. He tells of a calm mood with composed people, despite the many victims and the great devastation:

> The streets were full of people, recumbent, and some in sound sleep. They seemed to find a greater sense of security close to mother earth . . . The sidewalks and roadway were covered with fallen stones, wooden signs and the wreckage of brick walls . . . A changed San Francisco, indeed, from the secure, care-free, luxurious place of the day before![15]

He also tells how the disaster strengthened the sense of solidarity. The strong helped the weak and the old, and even egotistic people were transformed on that day. Why could it not always be like this, he continues – no poor, no rich, no envy? It was as if Christ walked the city and reigned over a willing people for a couple of days.

A somewhat less idyllic version of the disaster was penned by Rev. Joaquin Miller. He despised the city life of San Francisco and the tall buildings. In his opinion, earthquakes were perfectly innocuous. As a rule, more people die from stale fruit in the tropics than from earthquakes, he asserted. The fires, on the other hand, were a kind of purification.

> In truth I know nothing in nature quite so innocent as an earthquake ... I think that fires are good, especially in hell. San Francisco was not a clean city ... she was nasty ... From every corner you could see the flames bursting higher and higher from these costly stores which no city had ever had before, and the clouds for all three days and nights were most wonderful to behold.[16]

His dream that the city should be rebuilt with structures only two storeys high, though, was to remain unfulfilled.

The writer Jack London lived with his wife Charmian east of San Francisco and both of them registered the earthquake: 'We don't know but the Atlantic may be washing up at the feet of the Rocky Mountains',[17] he is reputed to have said to Charmian. Half an hour later, they were on horseback on their way to the city. From a vantage point in the mountains they saw smoke billowing up from San Francisco. After this, they travelled by train to Santa Rosa, which had also been badly hit, before arriving in San Francisco. Jack London walked among the ruins and flames all evening. 'I saw it all.' What he saw made such a strong impression on him that he was unable to write it down. 'I'll never write about this for anybody, no, I'll never write a word about it.' A very generous offer from a magazine in New York, however, proved too tempting for him, and on 5 May 1906 'The Story of an Eye-Witness Account' was published:

> Not in history has a modern imperial city been so completely destroyed. San Francisco is gone ... San Francisco, at the present time, is like the crater of a volcano, around which are camped tens of thousands of refugees.[18]

The Red Cross was responsible for relief work after the disaster. A considerable sum was collected from private individuals and from other US cities, but the authorities did not provide any emergency aid or money for reconstruction. Most of the money collected came from abroad. Fourteen countries donated a total of $474,000 – including a paltry $50 from Germany.[19] Reconstruction of the city was to become one of the world's largest construction projects, carried out by about 30,000 construction workers. It was not until the 1950s that law reform led to federal authorities in the USA being involved after natural disasters. We will see later just how they dealt with the biggest natural disaster in the USA in recent times – Hurricane Katrina.

Today, everyone is waiting for a new major and devastating quake from the San Andreas Fault. Geological surveys in the USA indicate that there is a 62 per cent chance of one or more earthquakes of more than 6.7 on the Richter scale occurring in the San Francisco region over the next 25 years.[20] No one knows what consequences this will have for San Francisco, but they will probably be considerable.

THE LAST CALL

'You-oo-oo gou-loo-loo come under the bloo-oo-oo boo-loo', an old lady cried, waving her arms. According to the *Los Angeles Daily Times*, a fanatical sect came into existence in April 1906. On the evening prior to the earthquake in San Francisco there had been 'wild scenes', with many people speaking in tongues.[21] During a prayer meeting on the day after the disaster, in South Main Street in Los Angeles, the floor suddenly began to shake under the feet of those taking part. Los Angeles was being hit by an aftershock from the San Andreas Fault – and many people fled out into the street. One of those present froze in fear. He later went home, but was convinced that God was ordering him to go to an evening meeting in Azusa Street. His name was Frank Bartleman.

Frank Bartleman, then 33 years old, had travelled to California in 1904 to run a mission station in Sacramento. This, however, did not go as well as planned, and he soon had to take part-time jobs in order to provide for his family. In December of the same year, they moved on to Los Angeles, where Bartleman was a preacher in the Methodist and Baptist communities. After the death of his three-year-old daughter Esther, he decided to devote his entire life to God and missionary activities. He prayed both day and night for a close, personal pact with God. His days were taken up with meetings, preaching sermons and the production of religious tracts. He distributed these himself, to banks and bars alike. He even tried to spread the gospel to the city brothels. In addition, he wrote articles for religious periodicals and was enormously productive.[22] Bartleman regarded his writing as a gift from God, given to him so he could create a religious revival. God was using him as his mouthpiece, he claimed. About a month before the earthquake in 1906, Bartleman had a vision in which God told him to write a new text with the title 'The Last Call'. He had taken part in a religious meeting in a dilapidated house in Bonnie Brae Street, where he met the preacher W. F. Seymour for the first time. Seymour was to become one of the most important figures in the Pentecostalist revival in California – an African-American who was blind in one eye, deeply religious and humble. In 'The Last Call', Bartleman stressed that God is prepared to give humanity one last chance to follow His word. Great things were about to come to pass, he believed.

> And now, once more, at the very end of the age, God calls. The last call, the midnight cry, is now upon us, sounding clearly in our ears. God will give this one more chance, the last. A final call, a worldwide revival. Then judgment upon the whole world. Some tremendous event is about to transpire.[23]

Three days before the earthquake in San Francisco, there was a great commotion after a meeting in the New Testament Church in Los Angeles. A woman had spoken in tongues, and people crowded around outside to discuss what this might mean. The prayer meetings had been taking place for some months, which could be a sign that Jesus was now revealing himself to the congregation. Bartleman was ordered by God to spend ten days in prayer. The problem was that he did not really know what God wanted.

Azusa Street lies in downtown Los Angeles, right next to the Asian business area. Today it is a small dead-end street of around a hundred metres. A small sign high up on a lamp-post soberly states the historical significance of the place. At the spot where the mission station once lay there is now a bank. As when Bartleman was alive, the area is a stamping ground for the very poorest in society, with soup kitchens and cheap rooms for long-term rent in the once fashionable and marble-clad downtown hotels from the 1920s.

When Bartleman heard of the disaster, he saw it in the light of his contact with God and was convinced that this was God's way of answering his prayers concerning a revival. Bartleman felt an enormous pressure on him to go to the evening meeting on 19 April. It had been moved from Bonnia Brae Street to 312 Azusa Street because of the large numbers. Bartleman had never been there before and he left home together with one Mother Wheaton. She had difficulty walking and Bartleman was impatient about waiting for her. 312 Azusa Street proved to be an dilapidated hovel with broken windows that had formerly been used by the Methodist church. The floor had been cleared of refuse and planks had been laid out over some boxes to serve as benches. There was still some time to go before there would be queues in the street to come in. By September that same year, it is estimated that 13,000 people would claim to have been saved by the sect.

On Sunday, 22 April Bartleman took 10,000 copies of the tract 'The Last Call' to the New Testament Church to be distributed. The following day, Bartleman claimed that he had once more been contacted by God, who had told him that he was to spread the word of the Lord. He soon finished a new tract called 'The Earthquake!!!' that was to be printed at once. Bartleman could feel God's anger with the population. 'San Francisco was a terribly wicked city.' Bartleman felt it was his personal task to inform people that it actually was God who had punished the inhabitants with the earthquake, before even worse things possibly would occur. He hurried to the printers with the message: each and every citizen was answerable to God. The tract was distributed in Los Angeles and San Diego. This called for a lot of courage. Bartleman tells about the difficult time he had and the reactions he met with:

> They were mad enough to kill me in some instances. Business was at a standstill after the news came from San Francisco. The people were paralysed with fear. This accounted to some extent for the influence of my tract. The pressure against me was terrific. All hell was surging around me to stop the message.

But I never faltered. I felt God's hand upon me continually in the matter. The people were appalled to see what God had to say about earthquakes.[24]

A few weeks later, the distribution of 'The Earthquake!!!' had been completed. 75,000 copies had been distributed in Los Angeles and the areas further to the south in the space of three weeks. At that time, Los Angeles was a city with just over 100,000 inhabitants. A further 50,000 were distributed in the San Francisco area. In the meantime, Bartleman had gone to meetings in Azusa Street, where people were both ecstatic and humble, singing and speaking in tongues. The Pentecostal Movement had been born.

'The Earthquake!!!' is a short, two-page document, the first page mainly consisting of biblical quotations.[25] Bartleman builds up his argument that earthquakes are God's punishment of mankind, that they are a direct consequence of God's wrath. He rhetorically asks what God has to do with earthquakes, and answers with a quote from the Bible that they 'are but the beginning of sorrows'. He then produces a prophecy made by the American preacher Lee Sprangler that, at a sign from Christ, a mighty religious revival will take place in the USA. Bartleman was a deeply religious person who, via 'The Earthquake!!!' sought to give a heartfelt warning of what could lie ahead if people turned their backs on God. He believed that there was a direct connection between the disaster and God. This must be seen in the context of the belief of evangelical movements in the Day of Judgment and a literal interpretation of the Bible. Towards the end of his tract, Bartleman gives his interpretation of the earthquake:

> Do you claim there is no God in earthquakes? – and yet you secretly curse Him in your hearts, for it. Be ware! And are you saying, 'we will build another city, a greater one?' Is it for God you would build it? Or for greater crime? *Remember Babel's tower!* . . . Now note the indisputable fact of history that it has ever been the wickedest cities that God has so visited.

The innocent will be spared, Bartleman continues, but the aftershock that hit Los Angeles the day after the main quake in San Francisco must be seen as a warning. 'Remember, while God has passed by, for the first time, He is still within hearing distance, and may return. Be careful of your speech.' Even if these were strong words, Bartleman went even further towards the end of the tract:

> The present, seeming calamity is but a blessing in disguise for a lost and ruined race, if we will have it so . . . Very soon this poor old earth will be struggling in the mighty throes of a final and complete dissolution. 'Be ye reconciled to God.'

Bartleman's gloomy doomsday prophecies almost send a shiver down the reader's spine – even a century later. Bartleman stated that the earthquake in San Francisco was God's warning to people along the Pacific coast, and that it had greatly assisted the revival taking place in Azusa Street. He was convinced that the disaster was God's work, and that God thus was indirectly behind the birth of the Pentecostal Movement. It is difficult, however, to gauge just how important the disaster was in that respect. For the time being, it must remain simply the interpretation of a profoundly religious and spiritual person. We do know, however, that the early Pentecostal Movement was highly preoccupied with the return of Christ, the millennium and the latter days, so many of its adherents may have felt as Bartleman did.

It was a widespread view in the early phase of the Pentecostal Movement that a revival would be followed by Christ's second coming.[26] According to the religious historian Allan Anderson, the speaking in tongues that characterized the religious meetings was interpreted as a sign that the end was nigh. Speaking in tongues would help the spread of the religious message all over the world, and would thus hasten Christ's second coming.

The 1906 earthquake was not the deciding factor for the birth of the movement. The seeds had already been sown some years earlier. But the message here is that disasters and human suffering can lead to changes in the worldview of those living at the time. Do people opt for what is secure and familiar? Do they allow themselves to be convinced by new leaders who appear on the scene? What is the social background of those most affected? Both Bartleman's accounts and the eschatological aspect of evangelical persuasions indicate that the California disaster must have made an enormous impression during those days in April 1906. The earthquake in San Francisco created conflicts between various interpretations of the relation between natural forces and the religious dimension. In this drama, there was an opening for changes and alternative views. It was impossible to remain unaffected by all this. The poor and the outcast in particular allowed themselves to be swept along. The Pentecostal Movement was met with revulsion by the established Protestants of Los Angeles.[27] The Pentecostalists' abandonment to the Spirit and hectic, energetic meetings were too much for most people. In addition, the movement had crossed a number of major barriers in American society at the time: blacks, poor people and women were key groups in the revival.

It can be enlightening to take a closer close at the relation between disasters, religious belief and poverty in the USA. Charleston in South Carolina had had a long history of storm devastation before a powerful earthquake shook the city in 1886. The quake led to considerable material damage, with 110 people losing their lives. Reactions to the earthquake were to divide the town in two. The whites wanted to normalize things and did not attach any deeper meaning to the quake. To the middle-class citizens, the reactions of the blacks were at least as frightening as the quake itself. Many blacks ran out into the streets shouting 'The Day of Judgment!'. Newspapers related how the blacks took over the public space – and to such a degree that the whites felt themselves threatened. The whites had a privileged position in the town, and the fifty per cent of the population who were black were poor and underprivileged.[28]

Aftershocks led to new waves of panic. Arguments and shouting filled the streets. Seven days after the quake, the situation was described as intolerable. Whites felt themselves banished to their ruined homes, while blacks held untamed 'religious orgies' in the centre of town. The historian Ted Steinberg claims that the disaster functioned as a pretext for the black population to reverse the fixed social norms of the town. The cause was mainly their interpretation of the earthquake as God's work. They were waiting for the Day of Judgment and were afraid of being punished. Geologists who came to the town to investigate the causes of the earthquake were indignant that the clergy did not share their scientific explanation. The reason was not that the blacks were ignorant or backward, as many whites claimed, but that they belonged to Baptist and Methodist communities. Such communities place more emphasis on the spiritual and have a more fundamentalist approach to the Bible. It is hardly a coincidence, then, that it was people with this religious and social background who came to dominate the Pentecostal Movement in 1906.

APOCALYPSE NOW?

What are the attitudes of the Pentecostal Movement today? Do members continue to believe that natural disasters can be a sign of the Day of Judgment?

The link between the Day of Judgment and natural disasters is mainly an American phenomenon. Details concerning the last days vary between the various Christian persuasions, but a central element in evangelical circles is that natural disasters have a role to play in the interpretations of when this time will come. This is partially based on a verse in the Bible: 'For nation shall rise against nation, and kingdom against kingdom: and there shall be earthquakes in diverse places, and there shall be famines and troubles: these are the beginnings of sorrows.'[29] On the Day of Judgment, people will be sent either to Heaven or Hell.

The Pentecostal Movement is the most rapidly expanding religion in the world. It has a following of perhaps as many as 500 million people, and growth is fastest in poor countries, especially in Asia and Latin America. It is also in the poorest countries that natural disasters hit hardest and affect most people, so that it is to be expected that the history of nature is often interpreted in a religious light. We will see later how both Christians and Moslems reacted to the tsunami disaster in Southeast Asia in 2004.

4

Natural Disasters in Metropolis

'The cities rise again.'
Rudyard Kipling

No major city has ever been so hard hit by an earthquake as on 28 July 1976. Not a single house in the city of Tangshan in China escaped undamaged – and most of them were razed to the ground. After the disaster, Tangshan was closed to foreigners for two whole years. Had it not been for the global network of seismographs, news of the disaster would probably never have leaked out of China. It was not until a year later that the authorities admitted that the earthquake had been the worst in 400 years. In the meantime, reconstruction had been planned at turbo speed. Officially, the earthquake claimed 240,000 lives – probably a substantial underestimate on the part of the authorities. Some people believe that as many as 800,000 of the million inhabitants of the city perished. No relief workers were allowed into China, however. This was based on the idea that it would insult the dignity of the people to accept help from outside. In addition, the region was extremely important for the country because of its industry and large natural resources. So the fastest reconstruction possible was thus an important objective. Supplies of food and relief workers were sent to Tangshan from the entire country. The process of reconstruction was to take ten years.[1]

Orders as to how the disaster was to be dealt with came from the top, from Mao Zedong, in the form of the campaign 'We must resist the earthquake and save ourselves'. A series of heroic accounts came from the ruins of the city, staged by the Communist Party. People who had lain trapped under buildings for days were said spontaneously to have uttered 'Long live the People's Liberation Army!' when pulled from the ruins and to have claimed that it was the thought of Mao that had enabled them to survive. Such stories were meant to be inspiring and instructive. They were models of how Chinese people ought to react to a disaster. Furthermore, the people were to learn to tackle adversity and 'swallow its bitterness'. By following Mao's advice, the Chinese people would survive the disaster.

Eyewitness accounts of the disaster and the course of events were conspicuously absent. The only published personal account came from a certain Qian Gang from the People's Liberation Army, which included the statement 'We cannot find words to express our gratitude to Party Chairman Mao and the Communist Party ... Earthquakes cannot subjugate a heroic people. We will continue to build Socialism with all our might.'

Shortly after the disaster, on 9 September, Mao died. Millions of people interpreted the earthquake as an omen of Mao's death, believing that he had attempted by the disaster to take as many people into the grave with him as possible.

The reconstruction of the city reflected the major political changes that were taking place in China. The struggle for power after Mao's death was won by more reformist forces, friendlier to impulses from abroad. Instead of rebuilding the city on the basis of Communist ideology, architects and urban planners were virtually given a free hand to make the new Tangshan earthquake resistant. Square city blocks and wide streets were to guarantee future evacuation – imitating the lessons learned from Lisbon in the eighteenth century. Tangshan became the symbol of the more reformist and 'new' China.[2]

The earthquake provides an insight into reactions to disasters in totalitarian countries, where authorities often refuse intervention from outside and impose a clamp-down on information about the extent of the disaster. A recent example of this is Burma after the tsunami disaster in December 2004. Very little information got out, and relief workers did not get in.

In China, restrictive politics and secrecy concerning natural disasters was to last for a long time. It was not until the autumn of 2005 that there were signals indicating that the extent of natural disasters should no longer be regarded as state secrets.[3] One of the reasons for this type of reaction is that relief from outside often has strings attached. Emergency aid is not always value neutral. Furthermore, disaster aid sends the message that the authorities are unable to manage crises on their own. Seen in this light, natural disasters can open up political negotiations and democratic overtures. After the earthquake in Bam (Iran) in December 2003, many people wondered if the disaster would have political consequences, especially after the USA offered emergency aid. It was an explicit hope in many people that disaster diplomacy and international relief work would soften up the relations between Iran and the West.

NATURAL DISASTERS AS OPPORTUNITIES

The location of Managua, the capital of Nicaragua, is extremely unfortunate. It is surrounded by active volcanoes. In addition, five earthquake zones pass through the city – and friction forces in 1972 unleashed a quake of 5.6 on the Richter scale. The author Eduardo Galeano wrote: 'The cathedral clock stops forever at the hour the earthquake lifts the city into the air. The quake shakes Managua and destroys it.'[4] Even though the earthquake was not particularly powerful, between 5,000 and 20,000 people died. Seventy per cent of the city's inhabitants were made homeless.[5]

For the first two days people had to fend for themselves, without any special help from the authorities or private organizations. All classes of society experienced the mental strain of the earthquake. A gigantic spontaneous evacuation started at once, with 250,000 people – more than half of the city's inhabitants – trying to get away to relations in neighbouring towns. The enormous class differences played a part in making the situation worse. The looting of warehouses, shops and homes began as soon as the dust had settled. Looting can be the only way of getting hold of goods when a society's big city structure collapses, although many people saw it as a chance to acquire material goods. The situation was out of control.[6] Nicaragua lacked evacuation plans and other necessary measures for dealing effectively with a disaster. The Nicaraguan authorities made the situation even worse: the country was controlled by a three-man junta.

If we go back in time, we can trace some of the country's problems to continual interference from the USA, which has sought to protect its own strategic interests in Latin America since the mid-nineteenth century. With support from the USA, Anastasio Somoza

García became the leader of the newly established national guard in 1933. The rebels, led by Augusto Sandino, accepted a peace agreement on condition that the USA left the country, but Sandino was murdered by the national guard in 1934. García continued to be supported by the USA and President Franklin D. Roosevelt: 'He may be a son of a bitch, but he's our son of a bitch.' This introduced a 43-year period of the Somoza family as leaders of the country.[7]

The 1972 earthquake exposed the anti-democratic and corrupt sides of the regime, which was then being led by Anastasio's son, 'Tachito' Somoza. The national guard itself took part in the looting and was the main seller of stolen goods on the black market. The situation was so bad that Somoza had to call in foreign troops to maintain order. The reconstruction of the city was taken over by Somoza and his people, with most of the international aid ending up in his lap. Moreover, much of the aid from the USA was given in order to save the dictatorship. The dominance of the Somoza family created a great deal of discontent among other politicians, who were looking for new alliances, as well as among the bourgeoisie, who complained about the lack of free competition. But the reconstruction of Managua initiated a boom that created a lot of jobs, with people flocking to the city. Two events were to prove pivotal as regards the country's future: a workers' strike in 1973 and a commando raid by the Sandinistas in 1974. Somoza was highly disliked among the working class, as described by Galeano: 'In the face of this catastrophe Tachito Somoza proves his virtues both as statesman and as businessman. He decrees that bricklayers shall work sixty hours a week without a centavo more in pay and declares: "This is the revolution of opportunities."'[8] The Sandinistas saw discontent with the government as an opportunity to bring down Somoza, and they managed to organize in support of their revolutionary programme. Somoza was forced onto the defensive, finally falling from power in 1979. A number of historians have claimed that the earthquake played a central role in this.[9]

Serious natural disasters in large cities have had political consequences. In a study of almost 90 disasters in the 1970s political problems were a frequent factor, mainly a result of an unwillingness on the part of the authorities to acknowledge the seriousness of the disasters and a tendency to interfere in an unfortunate way in the reconstruction work. The horizon for political leaders in underdeveloped countries can be a short one, the focus often being on quick solutions with maximum profit. Natural disasters only become a topic once they have taken place, disappearing from the agenda soon afterwards. An absence of democratic institutions can have serious consequences during long-term reconstructions. It is, however, important to avoid determinism: neither poverty nor undemocratic forms of government must inevitably exacerbate natural disasters when they occur.

Would the change of regime in Nicaragua have taken place without the earthquake? The disaster accelerated political processes that were already brewing in Nicaragua. The earthquake was the event that set the ball rolling.

BIG CITIES AND RISKS

A large number of the world's largest cities lie in the danger zone for earthquakes, floods, volcanic eruptions and hurricanes. Only about twenty mega-cities can be said to be completely safe from natural disasters. These figures are not in favour of the world's poor. About 90 per cent of risk cities lie in underdeveloped countries. The problems are greatest in Latin America, Africa and Asia. The geographer Mark Pelling claims that some of the blame can be placed on European colonization since the seventeenth century and the ensuing changes to patterns of settlement. One example of this is San Salvador, which continues to lie at the foot of an active volcano, even though the city has been reduced to ruins by earthquakes nine times since 1575.[10] These claims are supported by the anthropologist Anthony Oliver-Smith, who argues that the Spaniards established the town of Arequipa in the Andes mountains in 1540 without worrying as to why an area with good soil was so sparsely populated. The Indians knew that the steep mountains of the areas concealed a number of natural hazards that included landslides, earthquakes and volcanic eruptions. The Spaniards' lack of knowledge about the region was to cost many people their lives. In the seventeenth century alone the city was laid waste four times by earthquakes and once by a volcanic eruption.[11] Right up to the present day, Latin America suffers from a high vulnerability to natural disasters that partially derives from the early days of colonization.

There is another important cause in addition to imperialism: overpopulation. In poor countries, the number of floods and landslides has dramatically increased since the 1950s.[12] Urbanization is an important cause of the increase in disaster statistics. Global climate changes will reinforce the trend. Over half the world's population lives in towns, even though they only represent one per cent of the land surface of the earth. Even so, there has been a tendency for people to have overlooked and 'forgotten' nature when developing towns, the result of which has been that the weakest in society have had to suffer. In São Paulo in Brazil, for example, there were 220 floods and 180 landslides in 1996 alone that were the sole result of an overloading of the environment.[13] Almost two-thirds of all 'abnormal' deaths in major cities in Latin America and the Caribbean are due to natural disasters – and not to unrest or accidents.[14] The prospects are not good. Practically all increases in population in developing countries over the next decades will take place in cities, which will further increase the load imposed on nature in the marginal and vulnerable areas around the cities.[15] The slum areas in particular are exposed – and at present about 1 billion people live in such areas. This number will probably double over the next 15 years. It is important to remember, however, that poverty and urbanization do not by themselves lead to an increase in the number of natural disasters. The monster lurking in the depths is a lack of involvement in seeking to improve people's everyday safety.[16]

The risk of both natural disasters and other more long-term environmental problems in cities may take place as a result of a distorted power relationship between social classes and city districts. The fringe zones of major cities are often the most vulnerable and have suffered most under urban expansion. When vegetation becomes impoverished, this paves the way for soil erosion and the risk of slides and floods. As Pelling has pointed out, urbanization affects disasters in just as radical a way as disasters affect urbanization. When the cityscape changes, natural disasters will follow suit. So the risk of disasters in major cities is highly complex.

People's vulnerability depends among other things on social status. Ethnic minorities in the USA may be more vulnerable to natural disasters than whites because of lower income, poor housing, more exposed plots of land and the fact that they often lack any form of insurance.[17] In addition, people with low status take longer to get over mental problems and have a greater risk of a reduced standard of living if once hit. Ethnic background, religion and social status can also determine how people view risk. A study from California shows that African-Americans are more fatalistic about earthquakes than other groups. They feel there is little that can be done to protect them against new quakes. The group that is least concerned about natural disasters is white males.[18]

Natural disasters can expose social conflicts and problems in a society. The hurricane and flood in New Orleans in the USA in September 2005 showed the whole world how unequally natural disasters can affect a society with large class differences.

AN AMERICAN TRAGEDY

Everyone knew it would come. Everyone knew the effects would be devastating. No one was prepared for how unequally it would strike. Hurricane Katrina was one of the worst natural disasters in the history of the USA. At the same time, it was one of the most forewarned about. The results of Katrina further illustrate risk distribution in cites with a marked division of social strata.

Hurricane Katrina nucleated on 23 August 2005. After its conception off the coast of the Bahamas, the tropical low-pressure area rushed northwards towards Florida, where nine people lost their lives, before disappearing in the direction of the Mexican Gulf. At the same time, the wind increased in strength because of the unusually warm water of the Gulf. The wind velocity increased from 125 to 290 km/h. By this time, the hurricane had increased to Category 5, the maximum for a hurricane. Fortunately, Katrina lost some of its force before hitting the coast of Louisiana on the morning of 29 August. In addition, it missed New Orleans by 55 kilometres. But this was poor consolation as long as the hurricane left a trail of death and destruction in its wake. The hurricane brought a storm surge with it that can best be described as a tsunami. The levees that protected the city from the sea and the seawater bays were breached. The water flooded in over the low-lying city. 'Lake Bush', as the inhabitants referred to the collection of flood water, ended up covering 80 per cent of the city.

Poverty and vulnerability: George W. Bush consoling victims of Hurricane Katrina. He is ignoring the fact that the wealthy tend to be insured.

Almost a million people were evacuated in the days before Katrina hit the coast. But 80–100,000 people, many of them African-Americans, had no real possibility of leaving the city. Those who remained lacked cars or a place to go – and had neither the money nor resources needed for such an evacuation. In addition, many people stayed behind to guard the precious few possessions they had.[19]

The authorities feared that as many as 10,000 people had lost their lives. The figures were adjusted downwards in the weeks after the disaster, ending up at about 1,300 people. The photos of those left behind, of the despair, of armed soldiers and the collapse of a society, created a worldwide debate.

Was this really the USA in the year 2005? Had the federal authorities reacted too late? Four days after the hurricane New Orleans still lay under water while tens of thousands were living in uncertainty as to what was going to happen. No less than 230,000 square kilometres lying in four states were declared a disaster area. At the same time, Bush said at a press conference outside the city: 'Here's what I believe. I believe that the great city of New Orleans will rise again and be a greater city of New Orleans. (Applause.) I believe the town where I used to come from, Houston, Texas, to enjoy myself – occasionally too much – (laughter) – will be that very same town, that it will be a better place to come to. That's what I believe.'[20] Bush's credibility suffered a serious blow as he was apparently unaware of the well-documented risks connected to a powerful hurricane in the Mexican Gulf.

The history of hurricanes in Central America and the USA is long and painful. The greatest disaster in the history of the USA was probably linked to a hurricane and not to the earthquake in San Francisco. Galveston was hit by a tropical storm on 8 September 1900. At least 8,000 people lost their lives. The formerly flourishing city took a blow from which it never recovered. Several years before Katrina, it was well known that a storm could, at worst, inundate New Orleans and lead to the deaths of 80,000 to 100,000 people. Instead of strengthening the levees and protecting the city against a Category 5 hurricane, the Bush administration reduced the grants. One of the reasons was that the priorities of FEMA (the Federal Emergency Management Agency) changed after the Republicans took over power in 2001. Its newly appointed director, Joe M. Allbaugh, called disaster aid 'an oversized entitlement programme', for example. In his opinion, disaster aid ought to come from religious groups and the Salvation Army rather than from the authorities. With Michael Brown as director, the focus of FEMA switched yet more from natural disasters to terrorism. With the certainty that 100,000 people could possibly die as the result of a powerful hurricane, Brown stated: 'We were so ready for this. We planned for this kind of disaster for many years because we've always known about New Orleans.' But when the disaster had struck, Brown blamed the victims and said that the death toll was attributable to 'people who did not heed evacuation warnings'.[21]

For many people it came as a shock that the USA has millions of poor people who nevertheless have a lot to lose from natural disasters. A few commentators in the weeks after the disaster seemed to refuse to accept this as a fact. But no less than 37 million Americans live under the poverty line as it is defined today.[22] In large cities, the poor and underprivileged groups are often neglected – something that also makes them more vulnerable to danger. Normally, high risk in connection with natural forces is seen as a phenomenon that belongs to poor countries. Among the African-American population in the USA, the view was different. Two-thirds felt that the response to the disaster by the authorities would have been swifter if those remaining in New Orleans had been whites.[23] The disturbing scenes that took place in New Orleans for days after the hurricane also raised other questions. How would the USA be able to withstand possible future terror attacks when the country could not deal with a hurricane?

Compulsory evacuation from the flooded parts of the city continued during the week after the disaster. Everyone had to leave. With the enforced evacuation of poor people, a weight was also taken off the shoulders of the authorities. Mayor Ray Nagin pointed out that 'This city is for the first time free of drugs and violence, and we intend to keep it that way.'[24] Were we witnessing an ethnic cleansing of New Orleans, in front of the eyes of millions of TV watchers? 'We couldn't do it, but God did', a Republican politician said about the evacuation.[25] Three thousand jobs had disappeared in the city a month after the disaster.[26] There was speculation as to whether the poor inhabitants would ever return to their former neighbourhoods and a rumour that New Orleans was to be reconstructed as a capitalist utopia without any visible poverty.[27] Two and a half years after the disaster, it would seem that the opposite has taken place. The number of homeless had almost doubled in March 2008 compared with before Katrina. No city in the USA has ever had a larger proportion of homeless people than the 4 per cent of New Orleans.[28]

Hurricane Katrina had its scapegoats. President George W. Bush registered the worst opinion poll figures ever. Only 38 per cent felt he was doing a good job. Brown was kicked out of FEMA on 12 September 2005. It was not, however, just the local and federal authorities' unsatisfactory handling of the disaster that was criticized. If the war in Iraq had not swallowed so many resources, more soldiers would have been able to have been deployed in the relief work, many people claimed. This led to the handling of the national disaster being compared with the war in Iraq and the 'war against terrorism'. This criticism was categorically rejected by the president. Others claimed that USA's lack of involvement in the measures to reduce global warming was the cause. The hurricane was a punishment for not having ratified the Kyoto Accord. Nature hits back. Now the authorities could 'blame themselves'.

A newspaper in Indonesia made use of the chance to criticize the USA's foreign policy. 'The superpower United States has finally succumbed to nature's wrath. The US must eventually admit that it is unable to deal with the victims itself. Something has changed: Hurricane Katrina has destroyed some of the US's arrogance.'[29] The criticism took a more hostile turn in Iran. One of the leading figures in the Islamic Revolutionary Guard, Brigadier Masoud Jazayeri, claimed that Katrina showed it was possible to create a devastated warzone in any part of the USA: 'How could the White House, which is impotent in the face of a storm and a natural disaster, enter a military conflict with the powerful Islamic Republic of Iran, particularly with the precious experience that we gained in the eight-year war with Iraq?' America's leadership was like a balloon, Jazayeri claimed, that could easily puncture.[30] The president of Venezuela, Hugo Chavez, claimed for his part that the disaster revealed the depths of racial segregation in the USA, and that capitalism was the underlying cause.[31] Was Chavez right? Natural disasters are indicators of the fact that the relationship between man and the environment is not sustainable – and capitalist development *can* make the consequences greater. Despite the critical remarks from Jazayeri and Chavez, the disaster did not contribute to a change of diplomatic relations between the countries.[32] But the USA was caught napping by the offers of economic and practical aid that flooded in from the outside world. The superpower did not know what to do about the offer of 20 million barrels of oil from Iran, emergency aid from Europe – or 25,000 dollars from Sri Lanka.

The disaster quickly led to an upsurge of conservative religious attitudes – and not exclusively within Christian communities. Some Moslem groups in the USA also interpreted the disaster from a religious perspective. The leader of the Nation of Islam, Louis Farrakhan, saw a link between the ravages of the hurricane and the USA war in Iraq: 'For God's sake, see what we are suffering and see the suffering that we have visited on others. Stop it now before the God of Justice brings more natural disasters. You cannot arrest Him as a terrorist, but He is terrorizing America today.' He also pointed out that the hurricane only represents the beginning of all the problems that would follow: God would turn the forces of nature against the USA. 'It seems like it's pay day.'[33] Extremist Christians also had their opinions about the causes of the disaster. The group Columbia Christians for Life claimed that the purpose of the hurricane was to destroy five abortion clinics, or so-called 'child-murder-by-abortion-centers', in New Orleans, and that the hurricane looked like an embryo when seen from space. They expressed their view in a press release:

> Abortion (child-sacrifice) is a *national sin. 9-11* was a *national calamity*. Hurricane Katrina is certainly a regional calamity, and it has already had national effects. *repent america*. REPENT AND TURN TO THE ONLY ONE WHO CAN SAVE US, THE LORD JESUS CHRIST![34]

What lessons can be learned from Katrina? The hurricane's far from random 'selection' of victims in New Orleans make it easier to understand the dramatic differences in the extent of the natural disaster between rich and poor countries. Disasters never affect rich and poor in the same way. We can compare the vulnerability of Latin America to hurricanes with that of the USA: Hurricane Andrew resulted in 55 deaths in the USA in 1992, while material damage amounted to no less than $22 billion.[35] Over 130,000 homes were destroyed and almost 90,000 people lost their jobs. Contrast this with a corresponding hurricane in Central America, which set development back twenty years. Eleven thousand people died when enormous quantities of precipitation from Hurricane Mitch caused flooding and landslides in Honduras and Nicaragua. In the space of two days, it rained as much as in a year, and the force of the wind was particularly severe in the mountain areas. About a million landslides occurred on slopes throughout the country, primarily caused by overpopulation and deforestation. Hurricane Mitch was not nearly as devastating as Hurricane Andrew, but 11,000 people lost their lives even so. The trend is typical: material devastation in rich countries is greater in rich countries than in poor ones, but the death toll is considerably reduced in rich countries as a result of a better standard of living and a lesser degree of vulnerability. Mental reactions will also be more serious in societies where many people have lost family members, loved ones and children. A study from Nicaragua has shown that post-traumatic stress reactions were widespread after Mitch. The city of Posoltega was hard hit by the hurricane, and a questionnaire at a school revealed that 85 per cent of the children had either themselves been seriously injured or known someone who had been.[36] Many had lost those closest to them. A hatred of the authorities was strong among half of those asked – they felt let down because they had not been warned and because they received too little aid after the disaster. Depression was widespread.

Natural disasters can bring the weaknesses of society and the authorities' lack of priorities out into the light. Paralysed leaders, great social inequalities and poverty are part of the recipe for chaos and a devastation of the social order after a natural disaster. Hurricane Katrina showed that the USA must do more than just focus on terrorism in order to maintain security in its own country. A new powerful hurricane in the Mexican Gulf and a new and devastating earthquake in San Francisco will both come sooner or later. It is just a matter of time.

METROPOLIS IN RUINS

Big cities are like fortresses. People think they will last for ever – that they are invulnerable. But without continual maintenance they decay and crumble. Nature gradually takes over. Cities are often presented as symbols of our having taken control over nature. This, though, can be an illusion. As Mike Davis has pointed out, a constant battle is being fought to keep nature at bay from the cities. Normally, it advances by degrees, almost imperceptibly, although this process accelerates during natural disasters. Flood waters have to be kept out, rifts from earthquakes have to be sealed, bridges repaired, water purified, rubbish removed and scars from landslides replanted. This is an inevitable, never-ending task, Davis claims. Big cities, in an ecological context, are the most dramatic constructions that have ever been erected. Despite this, we know more about the ecology of rainforests than big cities, he continues: 'who has ever put a microscope to the ruins of Metropolis?'[37]

Jack London is one of those who has made use of fiction to tell the story of the decline and fall of the big city. In his 1912 book *The Scarlet Plague*, San Francisco is the setting for a narrative where nature has become just as bloody as the capitalist society it has destroyed. In the streets only the strongest survive. The decline is complete. The hero of the story is an academic who is trying to regain control over the savagery and re-establish a civilized society.[38] Disasters in Metropolis remind us that we will never remain invulnerable. No matter whether it is a question of natural disasters or terrorist attacks, it is not possible to keep threats and hazards at a distance. Barbarism lurks behind every house corner. After Katrina, the resulting possibility of savagery was what frightened Americans most. It is, however, important to remember that this view is partially determined by an conception where disasters are seen as a 'show' – as the disaster researcher David Alexander has pointed out.[39] Savagery and brutality are common elements in Hollywood presentations of disasters, where people are either heroes or villains. Typically, the heroes will appear, without any form of prior knowledge, and tame 'anarchy' by force alone. The comparison is relevant for New Orleans, where the military were deployed to enforce the rule of law and order. It must be said that New Orleans is obviously a city with a great deal of criminality. In relation to the population, ten times as many murders take place as in other American cities. Even so, it is difficult to know what actually takes place in a big city when disaster strikes. One is influenced by the view of the situation as portrayed by single individuals via the media coverage. One of the most important reminders of the fact that the 'Hollywood version' of Katrina was a construct was the media presentation of people left behind clearing a path through the deep flood waters, carrying food. While an African-American was described as a 'looter', two whites had simply 'found' food at a supermarket.

LONG LIVE CITIES!

Cities do not disappear. Not even devastating natural disasters or atomic bombs can compel the inhabitants or authorities to relinquish them. Hiroshima was rebuilt after the Second World War, as was Nagasaki. Cities are too important to be abandoned. There are examples of cities actually disappearing for good, but they are few and are way back in history. The cities of Pompeii and Herculaneum are perhaps the best known, both of which were buried under ash from the volcano Vesuvius in 79 AD. There are furthermore just over 40 cities that were abandoned after having been laid waste in the period from 1100 to 1800.[40] Over the past two centuries, on the other hand, cities have been reconstructed time after time. One of the causes could be the emergence of national states that better protect their urban interests and citizens. Cities are also important for a sense of belonging and identity for a great number of people. Are there general characteristics regarding the rebuilding of cities from ruins – and how long it takes?

A major study in the 1970s of how cities are influenced by natural disasters concluded that rebuilding after a disaster follows a distinct pattern and is predictable.[41] If that is true, we can say a great deal about the effects of disasters on cities and about the phases that follow in the wake of a hurricane or an earthquake. One of the most striking things is that the reconstruction period can be divided into four overlapping phases, each of which takes about ten times as long as the former one. The first phase focuses on emergency measures and starts as soon as the disaster happens. Survivors have to be saved and injured people helped. This is the crisis phase, where life and society have been changed in a dramatic way and usual activities have come to a sudden end. This phase can last from several days to several weeks, ending when the salvage operations and search for survivors is over. Both in San Francisco in 1906 and in New Orleans in 2005 this phase took about four weeks. At this point the most important transport arteries have been re-opened and it is possible to get underway with the second phase – normalization. Now what is still capable of being used in terms of infrastructure and buildings is made use of. The refugees return, roads are opened, food supplies ensured and the clearing-up work completed. Those who return to the city try to re-establish order in their homes, and many of the most important urban functions get going again. This phase can take anything from a few months to over a year. Whereas the characteristic of the first phase is the collapse of society's normal functions, the second phase is typified by their re-establishment. The two final phases are both characterized by comprehensive reconstruction. In the third phase the city is raised back to the level it was at prior to the disaster. Dwellings are built, working life returns to its normal rhythm and unemployment declines. The city experiences a 'building boom' and many new jobs are created. During the fourth and final phase, the construction projects are completed. For San Francisco after the earthquake of 1906 this phase took 20 years, although ten years is the usual period needed nowadays.

Even though there are general characteristics for how big cities react to natural disasters, it is nevertheless often the variations – what makes disasters differ from each other – that is the most interesting. The extent to which a disaster will have far-reaching political consequences as in Nicaragua in 1972, or expose problems of poverty as in New Orleans in 2005, will depend on local social and political conditions. One of the most important themes in disaster research over the last twenty years is vulnerability. The concept has increased an understanding of how people and societies are affected by a hazard. Vulnerability is a wide-ranging concept that can be divided into three strands. When people are exposed to a hazard, exposure will determine how hard they are hit at a certain location. People who live on a flood plain will be more exposed to floods than those living in higher land and will thus be more vulnerable. Resistance reflects how well people can stand up to the hazard to which they are exposed. This will depend on economic, mental and physical conditions. And finally there is the capacity to adjust to threats from the environment and to avoid harm, which will decided how long a society will have to toil at overcoming the damage inflicted on it.[42] As the disaster researcher Kenneth Hewitt has pointed out, the number of disasters in large cities will probably simply increase, depending on how they otherwise develop. This trend will last until dissimilarities leading to vulnerability are finally evened out and reduced.[43]

The reconstruction of a city does not only take place at the material level. In addition to the physical reconstruction there are more hidden processes of a political and ideological nature. The disaster must be remembered and there is a wish to mark the fact that the city has arisen once more as a better place and that further growth can be ensured. Lawrence Vale and Thomas Campanella carry out research into the ability of cities to recover from disasters. They claim that it is just as much a rhetorical and ideological knack as it is a practical objective. In this connection, we can recall President Bush's statements in the days after Hurricane Katrina. Even after the very biggest natural disasters, as in Tangshan, the devastation is interpreted as an opportunity for reform, and the authorities use this opportunity to influence the morale of the population. A good example comes from the time after the Kantō earthquake that shook Tokyo and Yokohama in September 1923. The earthquake claimed more than 100,000 lives and 40 per cent of Tokyo was laid waste, partially by fires that raged for days on end. The department of education initiated a large-scale collection of accounts from people who had survived the disaster, publishing them in three volumes that were rapidly distributed to Japanese schoolchildren. The publications were part of a strategy to improve moral values in a Japan that many felt was in a state of decline. Through stories of bravery, a spirit of self-sacrifice, humility and loyalty to the Emperor the authorities attempted to resurrect values they felt were under threat. Ever since the beginning of the twentieth century Japan had experienced political and social unrest. Koreans in Tokyo were accused of having been arsonists and having poisoned the drinking water, which exacerbated this and resulted in thousands of Koreans being killed in violent disturbances. So the authorities hoped that the stories of the disaster could prove a source of renewal and benefit the entire country.[44]

The reconstruction of a city, then, includes the construction of new narratives at both the political, psychological and symbolical level. Symbolic acts and the erecting of monuments are an important expression of a society's ability to return to normal life. Since an important function of authorities is to protect its citizens against hazards, major disasters in cities are a direct challenge to their competence and authority. The ability to recover is crucial in strengthening confidence in those in charge. The process can be unpredictable and complex – and weak groups are often excluded. The priorities of the authorities are clearly brought out into the open in this process, and discontent with efforts can have political consequences. After major disasters, the ability of a society to recover has a significance that goes beyond the areas laid waste. Recovery gains prestige at a national level, and local interests are mixed with national ones. This happened after Katrina, where President George W. Bush encouraged voluntary efforts and solidarity along the lines of 9/11: 'America will overcome this ordeal, and we will be stronger for it.'[45]

A possible conclusion is that the need to mark strength and vision in reconstructing a city leads to new urban plans with long-term goals for reducing new hazards. This is, however, not always the case. Reconstruction takes place based on ideas and values that have roots in the mode of thought of the established authorities. Radical proposals normally remain unheeded. The exception is when the disaster also leads to a change of regime. The major changes in both Lisbon and Tangshan took place precisely at times when new leaders with a vision of change took over. This resulted in new action plans, new architecture, new ideas – and a new deal.

5

Among the High Mountains and Deep Fjords

'My heart is as old as the earth.'
Hans Børli

Something dramatic happens to the topography of South Norway at the transition from east to west. The undulating, almost gentle hills are gradually replaced by a completely different landscape. If, for example, you drive northwest from Dombås towards Romsdalen, you can clearly see the changes. The further west you get, the steeper the sides of the mountains become. After the small town of Bjorli comes the point of no return. Against the horizon towers a dark, massive wall of rock – almost like a terminus. What lies ahead at the end of the valley? You are apparently driving directly towards a ravine, with vast walls of rock looming up on three sides. If you take a closer look at one of the mountain tops, you can just make out some slender power masts on the edge of a precipice. They are like specks of dust up there, and you cringe behind the wheel at the vastness of it all. If you let your gaze sink to the bottom of the valley, you can see farms clinging to sides that are full of scars and rubble from avalanches. At some points, the houses have been built in the shelter of huge rocks that were once part of the mountains. The valley swings sharply westwards. Romsdalshorn twists skywards and you can see Trollveggen with its sharp peaks. Keep going – soon you can breathe a sigh of relief at having got through one of the most dramatic landscapes Norway has to offer. Why here? Unlike in east Norway, the mountains of west Norway are off-balance. The imbalance has been created by glaciers that have dug out deep valleys and left behind sharp peaks and an alpine landscape. When the weight of the ice disappeared, the land gradually began to rise back up to its original level. This phenomenon is known as isostasis. And we are not talking about trifles here – parts of Norway and Scandinavia still continue to rise by eight millimetres a year. The Himalayas, by comparison, grow at a speed of about ten millimetres a year. When sections of the mountain fall down, it is as if nature is trying to redress the balance disturbed by the ice. As a result, this part of the country experiences the most landslides and the most natural disasters.

NORWEGIAN DISASTERS

Norway is reckoned to be a stable country in terms of nature: a country without volcanoes and earthquakes. It is situated far from the plate boundaries that are exposed to earthquakes. Geophysical natural disasters thus occupy only a small place in the Norwegian consciousness. Even though Norwegian natural disasters are numerous, they are small by international standards. The closest Norway came to a major disaster is when about 4,300 Swedish soldiers froze to death in the Tydal mountains in 1719. But these soldiers were at war, and the disaster is regarded more as a victory than a tragedy.

Natural disasters in Norway are mainly due to processes that affect the earth's surface. As many as 2,000 people have lost their lives in landslides since 1850 alone.[1] As late as 1983 a landslip in Sognefjorden caused a ten-metre-high tsunami.[2] An awareness of natural disasters and the danger of going on holidays in high-risk areas has probably increased since the Boxing Day tsunami of 2004. After the tsunami disaster a number of prominent Norwegian politicians soothingly stated that nothing similar would ever happen in Norway. These utterances gave rise to both surprise and indignation among west-country Norwegians who themselves had experienced natural disasters.

In addition to several scores of tsunamis in Norwegian fjords, the Norwegian coast was devastated by a wave about 8,000 years ago as a result of the Storegga underwater slide in the Norwegian Sea. The consequences for the early settlement along the coast – the so-called Fosna culture – must have been catastrophic. The wave was up to 11 metres high and it swept in for hundreds of metres over Nordvestlandet.[3] As far north as Tromsø the tsunami was four metres high. The risk of a new slide of the same dimensions is, on the other hand, small. Since the Storegga slide there have been thousands of avalanches, landslides, floods and tsunamis in Norway. A great many of these developed into disasters – also in modern times. What will the future bring? Geologists who work with slides claim that we must reckon on having seven to ten major landslides in Norway over the next 100 years.[4] Several of them may cause tsunamis – and the number of people who will lose their lives may be as high as 2,000 for the whole period.

During the winter half of the year Tafjord is a dark place where the sun is kept out by the surrounding mountains. It seems no less gloomy a place when one considers the natural disasters that have devastated the area. If one looks more closely between the wooden houses, one can see enormous boulders scattered around, as if tossed there by giants. The force behind this came from the mountains that shut out the sun. The houses rest on the deposits of former landslides. The probability of a new rockslide here is small, but a bit further out in the fjord an enormous block of stone is slowly but surely tipping outwards. The locality is one of about twenty-five in the searchlight of the Geological Survey of Norway (NGU). In the wake of the tsunami disaster in Southeast Asia in 2004 the Norwegian government granted a further NOK 15 million for the surveillance of unsafe sections of mountains in Møre and Romsdal. Monitoring fissures in mountain sides and making tsunami models are among the research activities.[5]

The worst thing about rockslides is not the destruction that can arise on the downpath of the rock, but the fact that they can create tsunamis. The height of the waves can be enormous and locally can thus be far more devastating than those that occur in the world's oceans. The highest recorded tsunami in idyllic fjord Norway is over 70 metres.[6] Because of the scattered settlement pattern, the death toll after disasters in the fjords has never risen to higher than 75 people. Even so, about 240 people have died as a result of tsunamis in Norway in historical times.[7] One of the worst of these struck Tafjord in 1934.

THE STRANGE DISASTER

Reverend Leonard Tafjord was lying in bed dozing. It was almost three in the morning on the night of 7 April 1934. He was sleeping

Funeral procession on the way to Dale church. A rockslide in Tafjord in 1934 had dramatic consequences: a tsunami of up to 64 metres in height took the lives of 40 people. At the water's edge the remains can be seen of six houses, smashed to smithereens.

on the second floor of his parents' small house at the far end of the Tafjord fjord and was at last beginning to fall asleep. He had been lying for a while in bed thinking of his childhood. Suddenly the house began to shake violently. Outside, a thunderous sound could be heard. It was as if all of Tafjord was full of drifting snow. Then there was total silence. The sounds disappeared as suddenly as they had come, and the vicar decided to go back to bed. All at once he heard a roar, and hurried back to the window. He did not know if he could believe what he saw:

> The white crests of the waves pass outside the window . . . In the gleam of light I can see the fjord seething – boats, beams, houses, everything. Then the transformer goes and everything turns pitch black . . . Then the noise returns, even more frightening than before. I hear a window break in the post office underneath – with a mighty din the water floods in over the floor, overturns the furniture. It must be the day of the Lord! I begin to pray in my distress, pray for a merciful death.[8]

A large section of the mountain along the Tafjord had succumbed to gravity and slid over 700 metres down into the fjord. As
much as two million cubic metres of rock had broken off. The
tsunami first came to the village of Fjørå, which lies two kilometres
from where the rockslide occurred. Three waves swept away practically all structures lining the water's edge and only two of the 17
who lost their lives were later found. At the same time, the wave
swept the shore clean the 5.5 km southwards to Tafjord, which was
hit by a wave ranging between 5 and 16 metres in height. In utter
darkness houses were crushed and boats flung hundreds of metres
inland. The *Corona* ended up far inland with its siren blaring, as if it
knew what was taking place. A few hundred metres from Tafjord's
house, a family of six had survived by seeking shelter on a huge
boulder. The water reached their chests when the wave came and,
in some miraculous way, they had managed to cling on. Many people ran up the valley away from the fjord for fear of further waves.
When dawn came, it was clear that 23 people from Tafjord had lost
their lives. The survivors were in a state of shock. Many of them
thought that the Day of Judgment had come. Reverend Tafjord
managed to inform the outside world of the disaster via telephone
– the five or six farms in Tafjord had no road connection. He stayed
at the telephone exchange and was contacted by people from both
at home and abroad. Bergen wanted to send food and clothing,
newspapers were eager for news – and in England they wanted to
know how many people had lost their lives. A funeral parlour wanted a deal on the sale of coffins, and an anonymous caller wanted to
know if a religious revival could come to the village after the disaster. Two planes with journalists landed the same day and quickly
sent reports off to Oslo.[9] Eyewitness accounts, facts and photographs immediately became part of everyday media nationwide.
'A terrible natural disaster that has cost 40 people their lives last
night struck Tafjord and Fjørå in Sunnmøre,' *Aftenposten* wrote. The
whole country was in a state of shock. Reactions were also strong in
England, which had friendly ties with Norway. The *Times* was one
of the newspapers that covered the disaster, expressing its deep
sympathy in a leader: 'So many British tourists visit the fiords in
summer that the strange and terrible disaster which has destroyed
two villages on the west coast of Norway will arouse far more than
a platonic sympathy in this country.'[10]

Volunteers from the entire region flocked to Tafjord to help in the clearing up operation. The fjord had to be cleared of flotsam, new roads built and the dead searched for. What was the attitude of the government to the disaster? Right after the crisis, the minister of social affairs said he felt that provisionally there was no need of aid from the authorities, since all those in the families affected had been killed, and those who had survived seemed to be managing well. A collection was unnecessary, the minister continued, as there was no serious distress. The following day, he changed his view somewhat, saying that economic aid for the area *was* necessary. He gave his moral support to an appeal set in motion by *Aftenposten*, but was unable to contribute any government funding because of 'State budget difficulties'. Money poured in to *Aftenposten*, however, from at home and abroad – like some humanitarian tsunami: 'The wave returns. No mortal wave this time, but one that soothes and helps. Let the whole country now show what it owes that small, industrious community out there in the west, where the daily struggle for existence was so hard, and where the closing scene was as fateful as in a play by Ibsen.'[11] The royal family, newspapers, business concerns and private individuals all gave willingly. Artists donated works to auctions – and one factory sent Wellington boots. As far away as in Johannesburg in South Africa a dance was organized to raise money for the victims. Not all the money, however, was distributed – and the Wellington boots eventually had to be paid for. The rest was placed into a fund to support the survivors and to prevent future disasters. This created much discontent among people and ended with a court case. The fund was dissolved in 1972.[12]

FROM DISASTERS TO WORLD LITERATURE

Researchers, tourists and writers have all been inspired and humbled by the natural forces at Sunnmøre for hundreds of years. At the centre of events lies Hellesylt. The town unwillingly became a focus of interest after the tsunami disaster in Southeast Asia of December 2004. The reason for this lies some way out in the fjord. About 800 metres above the surface of the water, a huge portion of the mountain is shifting at the rate of several centimetres a year. The locality is one of those being monitored by geologists, headed by Lars Blikra. A rockslide from Åkneset will, sooner or later, end up in the fjord and create a tsunami of several tens of metres.[13] The inhabitants of Hellesylt seemed to have learnt to live with this hazard, but after the disaster in Asia both journalists and film teams have visited the town to get people to say something about what it is like to live under the threat of a natural disaster. Maximum magnification of the crisis on the part of all those coming from outside has made a strong impact on the local population, who are now also worried about what effect the disaster scenario can have on tourism and business life, as the area is one of the biggest tourist attractions in Norway.

Vicar Svein Runde has lived in Hellesylt for over 20 years and is familiar with the attitude towards disasters of the people of the Sunnmøre area. His summer cabin also lies directly above the fjord from Åkneset and would be one of the first places to be hit by any eventual tsunami.

'The media want banner headlines that we are unwilling to give and that we do not want, either,' he says. The parish hall lies at the centre of the town and, along with the old people's home and the school, these are all buildings that could be hit if everything went wrong. Both Runde and the others in the town are worried.

> People here also reacted to an article in the local paper last winter. They had shaded in the part of the town that would be hit by a wave. It was a grim sight. People find it hard to believe that the dimensions of the possible slide can be that big. We are waiting for the final conclusion . . . A lot is at stake. Eighty children go to school here.[14]

The area has always been exposed to rockslides and landslides – and one of them has found a place in world literature. The year was 1862, and Henrik Ibsen, 34 years old, was on his way northwards from Stryn in a horse-drawn carriage. He was on a study trip financed by the university senate in Kristiania to collect folklore in Gudbrandsdalen and in western Norway. He had been travelling for about a week when he arrived in Hellesylt one day in July. He took lodgings at the inn right next to the waterfall, just a stone's throw below the church. Ibsen went round the town for several days. He was well dressed, but people thought he was a strange fellow and wondered what he was really doing there. He was unwilling to state his business, but some people suspected he was a 'free-thinker'. Ibsen had yet to make his breakthrough as a dramatist and presented himself as a student. He talked to many of the local residents, and noted down everything he saw. In addition, Ibsen took a long look at the impressive mountains. It is uncertain how long Ibsen stayed in Hellesylt, but he is known to have visited the vicarage and to have had conversations with the vicar, Ole Barman.[15] Ibsen also talked to the vicar's wife about avalanches and rockslides. Originally, the vicarage was to have being built down by the fjord, but because of the considerable risk of avalanches there, Barman chose to built it up on the grassy mountain side instead. Several years later, there was a slide where the vicarage had originally been planned.

What did Barman think about all the avalanches? He pointed out that even if the local people were not afraid of nature, they were nevertheless deeply influenced by the overhanging danger the mountains represented. Like today's vicar, Barman was also well aware of people's attitudes to the dramatic scenery around Hellesylt.

> The Sunelving thinks twice before he speaks or acts. This he has presumably learnt from nature. If he wants to save himself from danger in the narrow, deep valley, he must remember these words: 'I go in danger where I tread.' If he is down in the valley, he risks, according to the season, having avalanches, landslides or rockslides descend on him . . . It would seem obvious to assume that the natural conditions here must have the effect of making the local residents thoughtful and reserved.[16]

It may have been this reserve that Ibsen was trying to penetrate in summer 1862.

Ibsen's central figure of Brand is an idealistic clergyman on his way across the stormy mountains to his native town by the fjord. The period is mid-nineteenth century, and the town is facing a crisis. People have been hard hit by a famine. The reason for this is flooding, drought and failed harvests. Brand believes that the famine is a punishment from God. On the church green the vicar hands out food rations, while all around the dark mountains loom. The fjord is ugly and narrow. After a heroic crossing of the stormy fjord, Brand is approached by people from the town. They need a clergyman, a man of his calibre. He accepts. But Brand is a hard man. There is only the narrow path that leads to God, and he sacrifices both his son and his wife for his calling. He is unwilling to move out of the shadow of the mountains. The dark, threatening nature becomes a metaphor of his own dark mind – and the harsh weather symbolises the difficult task he has leading the local people. Brand has a new church built, but on the day it is to be consecrated, he realises that this is not the right place to seek God. He throws the key into the river and takes his congregation with him up into the mountains. They gradually lose heart and turn against Brand. They chase him, hurl stones after him and drive him further into the mountains. Brand continues his search for God in the high mountains. There he meets the wild girl Gerd, who believes he is Christ himself. Gerd feels she is been pursued by a hawk, and the shot she lets fly at the bird starts an avalanche that buries both of them.

Brand was to mark Ibsen's breakthrough, and his ideas for the play may well have taken shape during his trip to Vestlandet. The stories of slides from the area around Hellesylt are many – and a number of them have been passed on by clergymen. Hans Strøm was a clergyman and natural scientist who visited the area in the 1760s. He was fascinated by the harsh, hazardous nature of the area, with its steep mountain sides. Avalanches, rockslides and sudden squalls all made Hellesylt and the fjord a dangerous place, in Strøm's opinion. In addition, the slides left 'naked, ugly fissures' in the mountain sides.[17] Despite all the dangers, the farmers were at least free, he continued. Strøm was here contrasting the freedom of the fjord people with the poor Danish copyholders.[18]

In 1701 the curate Knud Harboe was caught by a tsunami on the fjord after a rockslide. He was on his way to Hellesylt church in a rowing boat. That same church was destroyed in 1727 by an avalanche, after which it was moved to a safer spot. The avalanche took with it two farms and 14 people. In other words, Henrik Ibsen had plenty of stories to choose from in Hellesylt. It was perhaps here that he also found his role models for the characters in *Brand*. The depiction of nature in the play differs considerably from the prevailing contemporary Romantic conceptions of mountains, fjords and farmers. The mountains are not beautiful but dark and threatening, and the encounter between mountains and people is not harmonious but dangerous. Even so, the mountains in *Brand* symbolize freedom and truth, although the local inhabitants would rather have the path of resistance down in the valley.[19] Freedom is gained by defying danger, by getting to the very top – to the open spaces on the mountain plateau.

> My lowland life has now been played out;
> Up here on the plateau is freedom and God,
> Below, all the others grope blindly.[20]

Hellesylt and the Sunnylvsfjord are not alone in representing both the beautiful and the dramatically hazardous. Several hundred communities in west Norway have been subject to avalanches, rockslides and tsunamis. The area hardest hit is Lodalen, a little south of Hellesylt.

COME TO ME

Loen was a popular destination for both Norwegian and English tourists in the late-nineteenth century. Many arrived in Loen by way of the Nordfjord in cruise vessels, continuing by carriage to the lake, Lovannet. From there, a trip by boat was necessary to arrive at the final objective: the mountains and the glaciers. Up to eight hundred people would visit the area on the busiest days – and the tourists would be transported by boat past the rockslide-threatened Ramnefjellet mountain. The tourist industry gave important extra income during the summer, while the hundred or so residents would be left to themselves as long as winter held Bødal and Nesdal in its grip. The houses lay close to each other in clusters, normally of 8–10 houses separated by narrow passages. The small fields lay spread out, and the farmers had to climb up in the mountains to gether grass in order for their animals to survive the winter.

There were frequent rockslides and avalanches from the mountains – the people were used to all the thundering and cracking noises. Rockslides had claimed lives in the area around Lovannet on a number of occasions, but the calamities had been small and life had gradually returned to normal. Times, however, were about to change. The first warning came in 1904, during an evangelical meeting in the lodge building in Bødal. An old woman told of a grim vision. She had seen a Christlike figure stand beneath the great Ramnefjell, right on the far side of Lovannet. The figure had stretched out his arms towards her and said 'Come to me'. After that, Ramnefjell had crashed down into the water. A year later, she had the same vision. On the following day, her vision became reality. A large section of Ramnefjell collapsed into the lake, causing a tsunami forty metres high. A total of 61 people in the small communities lost their lives, among them the old woman.[21] The boats around the fjord were crushed, as were 80 buildings, with more than 200 farm animals perishing. The news of the disaster in Lodalen reached the capital by telegram:

> The need and misery of the survivors said to be indescribable. The two Bødal farms of Raudi and Gjertun, which are undamaged as they lie higher up, have taken in the unfortunate men and women who have lost their loved ones in such a violent and terrible way. There are heart-rending scenes seen in the narrow rooms.

Aftenposten announced the disaster on its front page on 19 January, although it was negotiations with Sweden for an independent Norwegian consulate that was given pride of place. Norway was then united with Sweden, but gained its independence in 1905 after a period of intense political tug of war and military rearmament. Several days later, *Aftenposten* chose to use the disaster to make a union-political point. A poem of nine verses with the title '21st January 1905' adorned its front page. The poem is an account of the events in Loen, mixed with metaphors of freedom and Norway's future:

> A word of thanks to you, in whom we trust:
> lead flood-swelled waters in a path more fair,
> remove the stony idols grey as dust
> that from our Swedish brother on us stare!

Aftenposten hoped that a 'royal flood' could sweep away the most conservative Swedish unionist forces and relied on King Oscar not leading the Norwegian people in under the 'edge of the slide'.

The prime minister wanted to send help to Loen as soon as the news reached Kristiania, but ever though a strong gale was making rescue and salvage work difficult, he was informed that help from outside was not necessary as yet.[22] The authorities were subsequently to maintain vulnerability to natural disasters in the affected areas by only offering financial support to those prepared to continue operating their farms.[23] Those who wanted to move would have to manage on their own. Even before the disaster, there had been a fear of a rockslide from Ramnefjellet, without this resulting in either some form of securing of the mountain or an evacuation plan for the local inhabitants. After the fatal rockslide in 1905, geologists believed that there was little chance of new slides of the same magnitude. New houses were safe if they were built a bit further away from the water's edge, it was claimed. The local inhabitants were not reassured by this – and a number of them decided to emigrate to America. Several farmers in Bødal and Nesdal contacted the authorities as late as 1931 to gain permission to lower the water level of the lake. This would reduce the extent of a new tsunami. Nothing came of this, partly because it would be an eyesore and seriously harm tourism.[24] Throughout Vestlandet people were afraid of similar disasters, and the Geological Survey of Norway had their hands full examining mountains exposed to potential rockslides.

The tsunami in Tafjord in 1934 reawoke bad memories in Loen, where many people were afraid of a new major rockslide. One permanent resident, Mrs Myrhaug, expressed it in this way: 'Not one single night all year round, summer as well as winter, do I go to bed without praying to the Lord to spare our home and children from misfortune.'[25] Throughout the summer of 1936 there were daily rockslides from Ramnefjellet. On the night of 13 September, the unthinkable happened: a huge new block of the rock face plunged into Lovannet, causing the worst Norwegian tsunami disaster ever. The section that broke away was three times as large as that of 1905. The wave was also larger, and no fewer than 73 people lost their lives. This marked the beginning of the end for the communities along the banks of Lovannet. People were paralysed with shock.

One correspondent reported that 'there were no outer signs of grief, no tears, no moans. The calamity had been so overwhelming that the survivors seemed half stunned.'[26] Eleven people were injured and 16 farms swept away. Mrs Myrhaug lost her life, along with her husband and three daughters.

This time, society was better equipped to deal with crises. Health staff were sent from Oslo to Loen by plane, and the dead and injured were collected in the one house left standing in Bødal. One of the doctors from Oslo stated that conditions were grim and that he had never seen anything like it. One of the survivors from the 1936 disaster speaks about the despondency afterwards: 'Why should all this happen once again? Why should we be struck down? Why should innocent children with every hope for the future be maimed, wounded and killed?'[27] The vicar tried to provide a tentative answer to what had happened during the commemorative service for the dead on 16 September: 'The Lord has spoken to us. God speaks through all that has happened – and he speaks the words of salvation.' He did not address the question as to why God had created the disaster, since mankind was so much more guilty of cruel acts than God was. The bishop in Bjørgvin agreed with the interpretations. In his opinion, the disaster was a metaphor of what one day will come, and it was not strange, therefore, that many of the survivors believed that the rockslide represented the Day of Judgment. According to the bishop, the disaster was proof that Jesus was right when he said 'Heaven and earth shall pass away, but my word shall not pass away.'[28]

'There is hardly a country in Europe that has had so many slide-disasters as Norway. Time after time, houses and clusters of houses have been hit, many lives lost and assets acquired through generations destroyed. Despite this, nothing is done to prevent such calamities recurring. It is as if nobody really cared.'[29] These assertions were placed in *Aftenposten* after the disaster. Something had to be done about the natural disasters in west Norway. There was broad agreement that the farms along Lovannet ought not to be rebuilt – most of the topsoil had been washed away, in any case. This time, the geologists believed that a new rockslide from Ramnefjellet was inevitable.

Was moving to another area too dramatic a solution? The spokesman felt the best thing would be for everyone to move, but that this was hardly practical: 'What is to be done with them? If you destroy an ant-hill, the ants will rebuild on the same spot – and the same will probably take place here.'[30] Some of the survivors wanted to return, but after a further large rockslide on 9 November, these plans were shelved. The last major rockslide from Ramnefjellet took place in 1950, when a tsunami 15 metres high swept the shore of Lovannet clean. By that time, the communities of Nesdal and Bødal had already moved and nobody lost their lives. Today, Lovannet is so full of avalanche debris that a new rockslide from Ramnefjellet would not constitute any great risk.[31]

SUFFERINGS

What is known about the psychological consequences of natural disasters and exposure to risk in Norway? This question is highly relevant, as no less than ten per cent of the Norwegian population lives in areas exposed to rockslides, landslides or avalanches. How does a constant threat of natural disaster affect people's everyday lives? Can the risk of future slides cause mental suffering?

There are no studies of mental reactions in the wake of the major Norwegian natural disasters. We know precious little about what the survivors of the disasters in Loen and Tafjord may have gone through, even though one can get an inkling from the newspaper reports and interviews after the event. Within disaster psychology, it is generally stressed that the reactions of disaster victims occur in definite patterns that show many similarities. The phases gone through include the shock and the reactions to the actual disaster as well as the processing phase, which may take a long time after the disaster.[32] So we can also learn from studies of natural disasters made in other countries – especially those with a similar cultural background. There are differences in how people react to natural disasters and manmade disasters such as wars, terrorism or shipwreck. Natural disasters are often viewed as being inevitable and dictated by fate, whereas those that are manmade can unleash a hunt for scapegoats and create strong emotional reactions.

An investigation of the mental health of survivors after one of the biggest slides we know of shows that the long-term effects after disasters must not be underestimated. The huge 261-metre-high

Vajont dam in Northern Italy was to supply electric power to thousands and was one of the large engineering projects of the 1960s. What people did not know was that the damming up of the river and the construction of the dam led to major changes in the stability of the surrounding mountains. Small rock-falls from slopes were omens that were not taken seriously, and little was done to prevent what has later been seen as inevitable: a huge section of the mountain crashed down into the artificial lake behind the dam on 9 October 1963. Several hundred million cubic metres of rock hit the surface of the water, causing a wave that washed over the undamaged dam, down into the valley on the other side.[33] The village of Longarone was completely washed away, and many other settlements were destroyed. The wave left behind it a moonlike, highly polished landscape and 1,909 people died. The mental health of the survivors was not examined until 36 years after the event. The result was depressing. Of those invited to take part in the survey, 16 were still unwilling to talk about the disaster.[34] Mental traumas began immediately after the disaster and had still a hold on 10 of the 39 who finally took part. These traumas included a fear of rain and storms, the sound of running water, and a fear of the dark.

Mental traumas after accidents and disasters are called post-traumatic stress disorder (PTSD) and, among other things, are characterized by recurrent bad memories and dreams. People with PTSD normally want to avoid getting into situations that can be associated with their bad experiences. The stress disturbances can cause those affected a great deal of distress. Insomnia, exaggerated vigilance, fits of rage and concentration difficulties are common.[35] At worst, the consequences can be abuse of alcohol, violence and split families. Natural disasters often lead to stress disturbances that last longer than three years, although the first eighteen months are generally the worst. In many cases, the symptoms can become chronic and last for decades – as was the case after Vajont.

The long valley west of the Loen waters is known as Oldedal, and it is one of the most exposed areas to slides in present-day Norway. Every year, innumerable avalanches rush down the curved, steep mountain sides into the valley, which stretches in to the Briksdal glacier. These avalanches can be up to 600 metres wide. Over a thousand residents of the valley live in constant fear of the avalanches, but have learned to live with that fear. One of them has put it this way:

> Of course it's dangerous here, but we know how to take care of ourselves . . . although one thinks of course of the drifts . . . and the school bus . . . yes, all of us do – think if the youngsters had been taken . . . but we're not interested in moving either – for it's here we grew up. And the house is, so to speak, safe – we've got shuttering in the walls. Though the road ought to have been better safer . . . those in power are responsible here – but they haven't done anything about it so far.[36]

What is special about Oldedal is that the frequent avalanches occur in a well-populated area with busy traffic. On average, the main road is closed because of slides five times a year. In some years, this may be up to 25 times. What is known for certain is that slides have resulted in loss of life here since the eighteenth century, though no deaths have occurred since 1949. In a survey from 1996, 85 per cent of those asked said that they had experienced avalanches at close hand.[37] More than half of those asked felt threatened – and practically everyone feared that slides would take people's lives in the future. On the other hand, they felt that their own homes were safe. About half of those asked felt that ensuring that the roads were free was badly or very badly taken care of – and that the authorities had let them down in this respect. A man from Oldedal underlined this by saying that 'people should not have been living here. It's not right. And now people pay too little attention simply because we've been lucky so long.' Despite this, very few indeed said that they were prepared to move. It should be added that the assessment of risk for people in Oldedal was complex, depending on place of residence, income, civil status and whether or not child care was involved.[38] Psychologists have investigated how people in the risk area reacted to a constant threat of a natural disaster and to slides – and how this affected their lives. It transpired that over 70 per cent of those who replied felt a high degree of concern regarding both their own security and that of others. Almost half experienced nervousness and anxiety in their everyday lives. A few suffered from insomnia.[39] Here too, however, many declined to take part in the survey, and their reason for doing so is not clear. Only 90 out of 369 individuals agreed to take part. For this reason, it is difficult to gain an overall picture of how people are actually affected. The group least affected by the risk of a slide was the one that did not actively seek information about slides and did not try to avoid them either. It was people with a large need of information and with problems in dealing with their own emotional reactions that were most exposed to such forms of strain as anxiety. No less than 19 out of 68 individuals could be placed in this group.

After the Tafjord disaster in 1934 traumas and problems were swept under the carpet. The young people there were told to forget the disaster, even though seven of the school desks were empty when school re-started. The local teacher has stated that

> The local inhabitants [did] everything to forget that night and talked about it as little as possible. When school re-started, with seven less pupils and a class divided into two, I tried to get the youngsters to think of other things. We did not talk any more about it, stayed silent and got on with our work.[40]

It was not until fifty years later that the bad memories were talked about, when a historian began to collect personal experiences of the disaster. For many survivors this was actually the first time they discussed it. Similar accounts come from Loen after the 1936 disaster, where one of the survivors, Anders Bødal, has said that a lid was put on the bad feelings. 'That was the way people dealt with such things then. You had to tackle everything yourself, although people probably could have done with some crisis counselling back in the 1930s as well.'[41] Survivors have stated that not a tear was shed until three weeks after the disaster and that it was only then that people fully realised what had happened. The events were to be forgotten and repressed. Only over the last few decades has disaster psychology – where dealing with traumatic experiences is a central component – become a common, accepted discipline.

The denial of risk and traumas connected with the loss of family members can be traced back a long way. In writings from antiquity and in Norse sagas one can find descriptions of reactions that now would seem to resemble post-traumatic stress disorders.[42] Ole Barman's account of life in Hellesylt in the nineteenth century and people's view of risk show many resemblances to present-day risk-denial. People along the banks of the fjord lived completely without fear of any avalanches, in his opinion. In 1862 Barman paid a visit to an elderly married couple. They were asked if it was dangerous to live in the mountains.[43] Rasmus, the old man, replied: 'No, we are safe here, for just above the house there is a deep cleft. Everything that comes down from the mountain will be caught in it.' Barman then asked them if they had ever been exposed to danger. Rasmus answered that they had not. But down by the fjord in wintertime it could be dangerous. As a six-year-old, he had gone down there along with his father and his elder brother. 'While we were on our way down, the snow broke between us. I was left behind and I saw my father and brother fly through the air with the layer of snow down into the fjord. I never saw them again.' A few years after Barman's visit, both Rasmus and his wife perished in an avalanche that took the house with it. As Barman laconically concluded: 'The man should never have said "We are safe here".'

So we can assume that both experiences of avalanches and the risk of future ones leaves profound traces behind in local communities – and has done throughout history. Risk or not, the worst scenario must be to be buried alive. Norvald Standal experienced precisely this when he and two others set out in search of a comrade who had been caught in an avalanche from the glacier Molaupsfonna in west Norway in 1971. This resulted in them being buried in a new avalanche.

In just a few seconds the avalanche packed itself around me and I was constricted by wet, heavy snow. Everything was dark. And silent . . . I prayed for help where I lay under the snow. Prayed to God. We tend to do that here in the fjord when the situation gets serious. I felt it helped me as I lay there. I was strengthened by prayer. But I did not believe that I could be got out of there . . . I began to realise that I was going to die. The darkness was total. I was freezing. It became harder and harder to breathe.[44]

Safe from danger? Can the inhabitants of Honningsvåg feel safe from the forces of nature? What makes societies vulnerable?

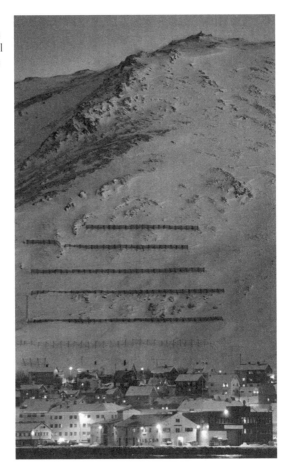

His two comrades did not make it, but Norvald Standal was rescued after having been buried for 18 hours.

Avalanches are common in alpine landscapes throughout the world, such as those in southern Europe, Scandinavia and America. They are normally set in motion by heavy snowfalls within a short space of time, which causes the snow to collapse, or a period of warm weather that causes the masses of snow to slide. The latter type of avalanche is often the bigger and more dangerous and, on steep mountain sides, can lead to avalanches that reach speeds of 200–300 km/h. Switzerland is one of the most exposed countries to avalanches, with an average of over 17,000 per year and an annual death rate of 24 people.[45] Humans can also bring about avalanches. So they have been actively used in war-fare. During the First World War, opposing troops from Italy and Austria-Hungary unleashed natural forces against each other. Record-high amounts of snow fell in the Alps in the winter of 1916–17, so it was possible to cause avalanches by, among other things, the use of artillery. The resulting avalanches rushed down the mountain sides, taking the lives of between 50,000 and 100,000 soldiers.[46]

One of the most dramatic slides we know of took place in Peru in 1970. An earthquake measuring 7.7 on the Richter scale hit a valley in the Andes mountains, triggering a large number of ava-lanches and mudflows (lahars). From one of the highest mountains in the area, the huge 6,750-metre-high El Huascarán, the avalanche bore with it loosely consolidated sediments, stones and water down towards the towns of Yungay and Ranrahirca at a speed of 100 km/h. People did not have a chance of escaping and both the towns were destroyed, resulting in the death of some 70,000 people. In addition, 2,000 villages and towns were devastated, with thousands more lives being lost and 140,000 people injured. About 6 per cent of Peru's inhabitants were made homeless. Even 30 years after the dis-aster, traces could still be found in its society. Much of the reason for this was the way the disaster was handled by the authorities. The American anthropologist Paul L. Doughty has pointed out that through the measures taken we have gained insight into the attitude of the authorities towards those affected. The most impor-tant assumptions and errors made were based on the idea that a central administration was the most effective, that local leaders are not competent to take important decisions (this must be left to the experts), that socio-cultural problems solve themselves and that the authorities out to stick to logistics and infrastructure, that help ought to be distributed in accordance with the local hier-archy, and that the poor need less help, since they had less to start with.[47]

The attitude of Norwegians to storms and disasters at sea dis-plays many similarities with that to natural disasters in the fjords. The underlying idea is of an omnipresent God and a nature that is outside human control. The mortality rate for fishermen in the nineteenth century was incredibly high – the sociologist Eilert Sundt found out that every fourth man in North Norway died at sea. The average lifespan in certain communities could be as low as 31 years. Sundt claimed that part of the reason for the high loss of human life at sea could be connected to human factors and class differences. It was most often the poor people who were affected, he claimed, because they had less seaworthy boats and had to put out to sea more frequently. The fishermen themselves had a different view of the loss of life. They were fatalistic and believed that God had something to do with it when the storm was at its worst and boats were smashed to smithereens. Such events were considered so far outside human control that the fishermen did not even feel it was worth learning how to swim. If the boat went down, there was a meaning behind it all.[48]

The worst catastrophe at sea we know from Norway is the so-called Titran disaster, when 140 men perished while fishing for herring. The Titran disaster from 1899 became part of the media's everyday coverage, where loss of human life shifted from being a local affair to something that concerned the nation as a whole. A record amount of money was collected for the bereaved.[49] And in the winter of 1849, a storm off Lofoten also resulted in considerable loss of life. Several hundred people may well have perished. Along the coast of Central Norway fatalism is a deeply ingrained way of life. After compulsory swimming lessons were introduced into the schools in the early 1950s a number of parents went on strike. Why should the young people learn to swim, since death by drowning was inevitable when the cutter sank? This attitude was widespread. Swimming lessons would only prolong the suffering of the fishermen, in their opinion.[50]

WARNINGS

Natural disasters have been warned about and discussed at both the prophetic and poetic levels in Norway. Prophecies of doom were commonplace in Norway in the late-nineteenth century, the main reason being the spread of evangelistic Christianity with its roots in the USA. The Christian revivalist movements came to Norway via missionaries in the 1870s and '80s. For both Jehovah's Witnesses and Adventists, the belief in a Day of Judgment and the establishing of a Millennium is seminal. Another characteristic of the revivalist movements is a literal interpretation of the biblical texts and a belief in miracles. Natural disasters become an aspect of the revivalist movements through – among other things – the gospels. On the basis of a literal interpretation of the Bible, wars, famine, plagues and earthquakes are signs of the latter days' arrival: 'For nation shall rise against nation, and kingdom against kingdom: and there shall be earthquakes in diverse places.'[51]

Among the prophecies linked to natural disasters, the vision of the so-called 'Man from Lebesby', Anton Johansen, was one of the most dramatic. From his native town in Lappland in Sweden he moved with his family to Lebesby in Finnmark at the age of 16. One autumn day ten years later, he had his first vision. Johansen saw Trondheim being destroyed in 1907. In the vision he was taken to various places where disasters would one day occur. Typical of Day-of-Judgment predictions is that they often predict a particular time or a particular event, such as an earthquake. Johansen saw before him a chart with a long list of years, next to which he saw the disasters. Trondheim was to be hit by a tsunami.[52]

> I was taken in the spirit to the area around Trondheim. I was standing on a shore looking out across the sea when the ground beneath my feet began to shake. The houses in Trondheim shook like aspen leaves, and a couple of tall wooden buildings fell down when the shore collapsed. Immediately afterwards, there was a huge booming sound out at sea, and a vast wave came rushing in at a furious speed and struck the rock wall . . .

Johansen also saw volcanic eruptions and hurricanes that would one day strike large portions of the world. Norway would be badly hit, but the mountains meant that the devastation caused by the hurricanes was considerably reduced. Johansen actively tried to prevent his prophecies being fulfilled.

> Along with thousands of others, I have prayed to the Lord for more than two years for these people to be spared such a calamity. That is why it has not yet taken place. And even though these calamities have not yet occurred, we must not forget to ask the Lord to spare us from them. For it seemed as if they absolutely were to come.

The Man from Lebesby was unwilling to talk in public about his visions, for fear of being scorned and derided. He himself felt a great inner strength and calm after that night, although he also had a feeling of a heavy responsibility.

Even though the prophecies were addressed to small religious groups, disasters like the one in Loen in 1905 have caused many other people to think of God as the Prime Mover. Dean Thorstein Gunnarsons is one of those who has studied Adventist movements in Norway, and his own admission from 1928 can still provide insight into what people think when encountering the dramatic: 'When natural forces seriously begin to manifest themselves, all of us become Adventists, we all say farewell to astronomy, meteorology, physics and geology, and revert to good old-fashioned theology.'[53]

Norway has also been destroyed in the literary world and, unlike the Man from Lebesby, Betrand Besigye has aimed at a broad and not necessarily religious audience. In his novel *The Swastika Star* from 2004, the main character Benjamin has a vision of how the earth is to be destroyed by natural disasters. Norway's capital, along with other major cities all over the world, does not escape this fate:

> I saw the terror earthquakes and the terror microbes invade Scandinavia. I saw the earthquakes create great fissures in the Nordic nations, chop up Norway and its capital, Oslo. I saw the quakes of death crush the inhabitants of Oslo with its buildings. I saw this people that never believed such a natural disaster could strike them be mown down like chickens for the slaughter. I saw this people that believed their own little place in the world was protected for all eternity from the wrath of mother earth, that believed such massacres were a foreign phenomenon, that believed that sort of calamity did not have anything to do with *them*, buried beneath the piles of cement!

Why will all this happen? According to Benjamin, humanity will finally have to pay for the atrocities it has committed against Mother Nature, for all pollution and the adopted role of ruler. Nature replies with an ethnic cleansing of the human race. Even the self-righteous Norwegians are included. Nature is brought forwards as an independent, dynamic entity beyond human control – on a par with an omnipotent God. The result is the ultimate apocalypse – one that makes the prophecies of the Man from Lebesby mere Sunday School reading.

6

Problem Children

'Day became night and light darkness.'
Cassius Dio, describing the eruption of Vesuvius in AD 79

In the course of hundreds of thousands of years, molten stone had slowly but steadily accumulated at a depth of several kilometres below the Santa María volcano in Guatemala. As the magma gradually rose towards the surface, the underground pressure also increased dramatically, and activity in the old volcanic crater changed radically. Clouds of steam and carbon dioxide rose up and hot springs started to spew out sulphur. In October 1902 the earth's crust was no longer able to withstand the enormous pressure from the magma chamber. The volcano split, with magma at a temperature of more than 1000°C bursting out of the magma chamber and up into the conduit of the volcano, pulverizing before everything shot out with enormous force. It was one of the most powerful eruptions on record. It could be heard all the way to Costa Rica, about 850 km away, and it left behind a fissure in the side of the mountain that was a kilometre across. A column of volcanic ash rose several tens of kilometres above the crater before it collapsed and formed a deadly carpet that spread out over the surrounding towns. At the same time, a strange sight could be observed in the town of Quetzaltenango, a few kilometres north of the volcano.

It was midday, yet as black as night because of the fall of ash, with the glowing fissure of Santa María the sole source of light. The Saturday market in Quetzaltenango had not taken place, so there were no stalls to be seen. But there was something else than vegetables and eager vendors that vied for people's attention that Saturday. A band with drums and trumpets was playing as loud as it could to drown out the sound of the roaring volcano. Much ash was falling, and volcanic bombs were smashing into the roofs of the surrounding houses. At the same time, more than a hundred local villages were in the process of being wiped out by streams of ash and slides. From time to time, earthquakes wrought havoc. What was the band doing here on a day like this? Accompanied by the volcano, a town crier lifted his lamp and shouted desperately the proclamation that had been signed by the country's president: *There are no active volcanoes in Guatemala.*

The government had decided that reality did not exist.

The town crier continued by saying that *if* anyone should happen to feel an earthquake, it was sure to have come from Mexico and not Guatemala, which was a calm country under the protection of its good president, Manuel Estrada Cabrera. The situation in the country was therefore normal, the town crier continued. Nothing would prevent the festival being celebrated in the capital from being a success.[1] At the same time, a large number of people in the mountains around the volcano were losing their lives. Nobody knows just how many, but it may well have been in excess of 5,000. Quetzaltenango was subsequently buried under half a metre of volcanic ash.

The time of the eruption did not suit the president at all. The outbreak started on the public holiday in honour of the Roman goddess Minerva – recently made the national saint of progress by the president. The day was just as much President Estrada Cabrera's homage to himself and it was being celebrated in the capital with wealthy guests from all round the world. The American president, Theodore Roosevelt, was among those wishing Estrada Cabrera every success with building up the country. But the Guatemalan president's thoughts were not with the people. It was to transpire that the education system in Guatemala was virtually non-existent. The author Eduardo Galeano has acidly described Estrada Cabrera's vision: 'He has conferred on himself the title of Educator of Peoples and Protector of Studious Youth, and in homage to himself celebrates each year the colossal feast for the goddess Minerva. In his Parthenon here, a full-scale replica of the Greek original, poets pluck their lyres as they announce that Guatemala City, the Athens of the New World, has a Pericles.'[2]

A study group from Chile that visited the country some time later found neither teachers, school materials, students nor school buildings. There was quite simply nothing to visit or look at. Instead of building schools for all the illiterate people in the country, the president had given priority to monuments praising European culture. But this is only one part of the story. The denial of the volcanic eruption was to the highest degree politically and economically motivated. The president wanted to try to convince the visitors and investors within the coffee industry that the country was a safe and secure place.[3] There *were* no volcanoes that could destroy the crops. Coffee production was the backbone of the economy; at the same time, people were starving and forced to work under terrible conditions. The lush ridges and the fertile soil around the volcanoes were the ideal place for cultivating coffee, as long as the volcanoes were dormant and access to cheap labour plentiful. President Estrada Cabrera did not get the truce from natural forces that he had hoped for – most of the coffee industry in the country was destroyed by the eruption. Santa María's intense eruption lasted for twenty hours, as it spewed out at least 20 cubic kilometres of volcanic particles (around the volume of 30 standard single dwellings every second). Santa María was active until 1913. Nine years later, a new eruption started, and the volcano has not yet become dormant.[4]

Latin America has a total of about 270 active volcanoes and is one of the areas in the world most exposed to volcanic hazards.[5] Santa María lies in a glowing belt of volcanoes that stretches right round the Pacific and is rightly known as 'the ring of fire'. Vulcanism has arisen as a result of the earth's crust beneath the Pacific being pressed down into the depths beneath South and Central America, where it melts because of the high temperature. The small drops of magma collect when they rise upwards in the earth's crust, gradually forming large accumulations, or magma chambers. This collision has been going on for about two hundred million years, and the enormous friction leads to numerous earthquakes in the same area. In Guatemala these are so frequent and powerful that the authorities have given up trying to get the railway system to work. Time after time, the rails get twisted and ruined. The earthquakes often give rise to landslides, a phenomenon that is intensified in the periods when this part of the world is hit by tropical storms.

The Pacaya volcano in Guatemala during an eruption in *1755*, as portrayed by the artist Petapa-Escuintla.

FIVE TOWNS. FIVE REACTIONS. ONE VOLCANO

20 February 1943. Dionisio Pulido was on his way out to the maize fields outside the town of Parícutin, in Mexico, one afternoon to set fire to a heap of branches when he saw the crack. On a small ridge he could see a fissure that was about half a metre deep. It had not been there previously. Dionisio did not examine the matter more closely and returned to his heap of branches. Suddenly, there was an earthquake, and it was then he saw it: ash and gas streamed out of the fissure that, within a short space of time, formed a two-metre-high smoking heap. Dionisio could hardly believe his own eyes.

> I was so stunned I hardly knew what to do . . . or what to think . . . and I couldn't find my wife, or my son, or my animals. At last I came to my senses and I remembered the sacred Lord of the Miracles. I shouted out 'Blessed Lord of the Miracles, you brought me into this world – now save me!' . . . I looked into the fissure where the smoke was rising and my fear disappeared for the first time.[6]

Sulphurous vapours covered the area. The volcano Parícutin had been born. It was to prove to be something of a problem child.

Around midnight, activity increased and after twenty-four hours the heap had grown into a 50-metre-high volcano. After a week, its size had once more doubled. The volcano was a so-called 'cinder cone', consisting of a fusion of volcanic fragments and volcanic glass that had been flung into the air. Ash and volcanic bombs fell onto the surrounding towns for a long time, gradually burying Dionisio's home. In March, the intensity increased, with a column of ash several kilometres high being spewed out. The volcanic bombs that were flung out were up to a metre in diameter and still hot. The volcano was adult after only a few months' existence. The evacuation of Parícutin began in June. San Juan was vacated a couple of months later. The lava streamed out of the flanks of the volcano, gradually placing a black lid over the towns. The only sign that there had once been human habitation there was a church tower that protruded from the smoking lava.

For the first time in history, researchers were able to study the birth, development and death of a volcano. For the local people it was not quite as exciting. Their crops were ruined and the forest died. The environmental consequences for the area around the volcano were disastrous. A famine was averted in 1943 thanks to national and international aid. Meanwhile, curious onlookers streamed to the area to see the volcano and talk to the local population. The outbreak was to last for nine years and destroy dwellings, towns and farmland before the volcano itself died in 1952. Returning to a normal way of life was to prove a long, costly affair for the Mexicans.

Prior to the eruption, the five towns around the volcano resembled each other as regards culture and economy. All of them had preserved their traditional Indian culture and language and were fairly isolated from what was taking place in the rest of Mexico. For this reason, one might assume that the towns adapted to their circumstances in roughly the same way over the following decades. That was not what happened. A study of the towns' development over a 30-year period by the sociologist Mary Lee Nolan has shown that the volcanic eruption set processes in motion that intensified and accelerated the cultural and social changes in the region.

The town of Parícutin was completely destroyed, and all its inhabitants, including Dionisio and his family, were moved to a new rural area 20 km away, known as Caltzontin. Parícutin had been more open to impulses from outside than several of the other towns. In addition, it had a strong farmers' movement. But because of conflicts with San Juan and what they regarded as a hostile and unchristian agricultural movement, the town was blamed for the eruption. Moreover, a number of blasphemous prophecies seem to have flourished in the period prior to the eruption. God's wrath had been roused by the people's sins, it was claimed. Thirty years later the town was still extremely poor and had also developed a class division between the farmers and the workers, who were newcomers. Furthermore, many people felt that the new name – Caltzontin – added to the alienation felt in the new location.

Angahuan was not completely destroyed, so it was not abandoned. The town was more traditional than the four others, with most of its inhabitants being indigenous peoples. Contact with the outside world was insignificant and sporadic and there were practically no new arrivals. This was to last until after the 1970s, with the entire community becoming increasingly marginalized and tradition-bound.

Zacan was ethnically more mixed than the other communities, with major changes resulting from the eruption. The town was not moved and farming was soon resumed. But many of the young people, especially those with an education, moved to other parts of Mexico or to the USA. It was on the whole the old people who remained behind – and they lived to a great extent off the money sent home by their relations.

Zirosto was more socially divided than the other communities prior to the eruption, with an emerging middle class that had more money and better land. This was to lead to complete social fragmentation between the indigenous peoples and the more prosperous Spanish descendants after the eruption. There was a struggle for the few resources available. The population was moved to three different locations and the identity of the town was lost for good.

San Juan Parangaricutiro was the most important and largest town as well as the seat of local administration. It was also the only one with electricity and telephones. Ethnically, it was divided between Spanish speakers and indigenous peoples, although the church had a well-known relic – a crucifix – that strengthened the feeling of unity in the town and had made it a place of pilgrimage. The town was buried under the streams of lava and most of the inhabitants moved to a new place they had selected for themselves. Many, though, refused to leave their farms and the church. This finally resulted in Bishop Luís Gómez starting a procession, with the crucifix out front, on the day the lava flows got as far as the church. The procession grew bigger and bigger on its way to the new town. Gradually, the crowd grew so dense that a chain of men had to be set up to make sure that the crucifix could pass. One of the places the procession passed was decorated, and people shouted 'Long live Jesus Christ' as the crucifix passed. Hope rose among the people of San Juan, and they grew convinced that God was leading them along the right road to their new home. The new San Juan was gradually modernized, but preserved its identity, as people had participated in deciding where it was to lie and by themselves taking part in its resurrection. The eruption and the removal strengthened this sense of unity.[7]

An important lesson from the Parícutin eruption is that communities, even at the local level, have various responses and reactions to natural disasters. It was the small differences between the towns that widened after the eruption and the evacuation. The unique responses to the eruption reflected the existing social pattern. This means that it is necessary to have a profound understanding of a society in order to know why it reacts as it does during a crisis.

MONITORING AND SCAPEGOATS

Few of the over 1,500 active volcanoes around the world are actively monitored. The reasons for this often have to do with a lack of funding and with the authorities in the countries concerned having too many other things that have high priority. Most of the active volcanoes lie in developing and poor countries. Colombia, for example, has 13 active volcanoes, Guatemala has 25, Chile 79 and Indonesia no less than 82. In Indonesia, as many as 155 volcanic eruptions are known to have influenced the population in recent times. At the same time, these countries are hit much harder by disasters than developed countries. In developing countries, costs related to natural disasters can be as high as 40 per cent of the GNP.[8] The explanation for this is the high vulnerability of these countries to natural disasters (and their low GNP). Furthermore, 99 per cent of those who have lost their lives during volcanic eruptions over the last few centuries have come from developing countries. This amounts to just over 260,000 people.[9]

As a result of the lack of measures taken and the high death figures, it is easy – in the wake of the disaster – to blame the authorities – or researchers. As in the Middle Ages, when the Jews were often blamed for natural disasters, it also helps nowadays to have someone to criticize. That makes the disasters easier to understand and accept. After the earthquake in Turkey in August 1999 corrupt contractors were blamed for the disaster because they had ignored construction regulations in order to save money. The same occurred in Italy and elsewhere, most recently in China in 2008.

Mexico has 35 volcanoes that have been active in recent times, without the country – until 1982 – having any programme to predict and warn of eruptions, any evacuation plans or anyone responsible for leading a possible crisis situation. Mexico's volcanoes were like ticking bombs in more than one sense, with major eruptions capable of having fatal consequences. This is precisely what happened in the 1980s. After an eruption from El Chichón in Mexico in March 1982, between 2,000 and 3,000 people died from the streams of glowing volcanic particles. The warnings were not taken seriously and nobody was prepared for the eruption. The state of alert was just as bad as in Colombia. A volcanic eruption from Nevado del Ruiz in 1985 exposed the authorities' unreadiness and is a useful example of the difficulties involved in dealing with natural disasters and risks in poor countries.

With its 5,389-metre peak, Nevado del Ruiz is a favourite spot for mountain climbers and hikers, despite it being an active volcano. One November day in 1984 moutaineers reported having seen clouds of sulphurous steam rising from the ice-covered mountain. It seemed as if it had woken up from centuries of sleep. An energy company with local interests and owners of the region's coffee plantations got together to find someone who could investigate what was going on. Considerable economic interests were at stake. But there was not a single geologist to be found in Colombia who knew anything about volcanoes and their eruptions, despite the fact that the country had 13 active volcanoes. Three young researchers got the job, one which they soon realized they would be unable to manage on their own: they were too inexperienced. Furthermore, several of the institutions involved argued about who had the overall responsibility, while all of them lacked money and competence. Finally, the geologists had to ask the UN for assistance. An international team was quickly commissioned to assist the Colombians in monitoring Nevado del Ruiz.

The ensuing intrigues and confusion have been clearly mapped. Both the geologist Barry Voigt and the journalist Victoria Bruce have shown how national problems, personal conflicts and attitudes all helped to influence the understanding of the volcano and the risk involved. Would it explode? Opinions were divided. As a forewarning, the volcano had a small eruption in September 1985 that sent ash over the surrounding towns. The population could expect more if the eruptions continued: the greatest danger was that an eruption could melt the glacier that covered the volcano. This would result in a stream of water, volcanic ash and other material accumulated on its downward course – a so-called lahar. On 7 October 1985 one of the surveys was complete. Its conclusion was that an eruption could wipe out the town of Armero, which lies exposed in a valley that stretches up towards the volcano. At best, the inhabitants would have two hours to evacuate the town.

Lava flows do not normally take lives. They often flow at walking pace down the flanks of volcanoes, making it easy for people to avoid them. It is the explosive eruptions that are dangerous, especially where streams of glowing particles and red-hot gas are formed. The speed of these pyroclastic flows can exceed 300 km/h. Such a stream took the lives of 30,000 people on the island of Martinique on 8 May 1902. A high gas content in the viscous magma makes the eruption explosive. In addition, lahars are a threat if volcanoes are covered by glaciers – as are Nevado del Ruiz and over 50 other volcanoes in Latin America. One of the best-known volcanic eruptions through a glacier took place on Iceland in 1995 under Vatnajøkul, which partially melted. Explosions can also cause large sections of volcanoes to collapse. If they lie in the vicinity of the coast or on islands, this can result in tsunamis. This is what happened, for example, with Krakatau in Indonesia in 1883, when 36,000 people were washed away by a tsunami up to 40 metres in height.

On the afternoon of 13 November 1985 two young researchers, Fernando Gil and Bernardo Salazar, were on their way to collect data from the seismic stations around Nevado de Ruiz. The seismographs can detect even small tremors in the ground and can therefore provide information about the risk of an impending eruption. On that day, one of the detectors placed a few kilometres northwest of the volcano had a very high reading. The problem was that no one knew how the data should be interpreted. Normally, all the information was collected and driven to the regional capital for analysis. Return information was rarely received. If Gil and Salazar had been able to interpret the signals that afternoon, they would have been scared out of their wits.

Nevado del Ruiz had erupted an hour and twenty minutes earlier. Ash and gas had already formed a 28-kilometre-high column above the volcano, but the cloud cover prevented the eruption from being discovered for several hours. Rain and ash fell onto several towns, and eventually the civil defence was placed on alert. Everyone knew what had happened. The rain died down as evening came and the air no longer contained any ash. The priest in Armero told the inhabitants of the town over a loudspeaker system that there had only been a minor eruption and that the danger was now over. People could put their trust in God. Most people were reassured by the priest's words.

Later that evening, Gil and Salazar were in the mountains. They had planned spending the night in a cabin when they heard an ear-splitting explosion. Then two more. Nevado del Ruiz had had three explosive eruptions – this time it was serious. While the volcanic bombs fell, Salazar phoned the headquarters of the energy company and told them to pass on an evacuation order to the civil defence. Gil was strongly influenced by what he saw: 'I felt like it was God reaching inside me, touching my soul.'[10] He fell to his knees and began to cry. A few minutes later, the ground started to shake and a wall of noise came down the valley in their direction. They fled in panic when they realized that the sounds came from a lahar that was growing with every second. The volcanic eruption had melted part of the glacier at the top of the mountain and the lahar was heading towards Armero. But no one in the town had been given the order to evacuate, and over the loud-speakers the priest was once more reassuring the population. The mayor repeated over local radio that there was no danger. Mean-while, the wave had grown into a 30–40-metre-high wall moving at a speed of ten metres a second down towards the valley, towards Armero. On the northern side of the volcano, a similar lahar en-gulfed the town of Chinchiná and took the lives of over a thousand people in the space of a minute.

Armero finally got the order to evacuate, although the meas-ures adopted were paltry, to put it mildly. A fireman went round the town, knocking on people's doors and trying to get them to flee. But they refused, since they had heard that there was not any danger. Electricity was down, as was the radio network. No one was prepared to listen to the fireman, who finally had to get out of town fast himself. At the same time, the mayor had radio contact with the civil defence, which gave him a final warning – which he chose to ignore. The last thing he reputedly said was: 'Wait a minute, I think Armero is being flooded'.[11] A large part of Armero was swept away by the devastating force of the large masses. Twenty-three thousand people died. Thousands of people needed medical help and the consequences for the economy and agri-culture were enormous. Large areas of land were ruined, 12,000 farm animals perished and 7,700 people were made homeless.[12]

The large loss of human life after the Nevado del Ruiz erup-tion is to a great extent due to social and human factors. It was not the forces of nature alone that can be blamed. It was known well in advance that the volcano would one day erupt and what conse-quences that could have, even though researchers disagreed about some of the interpretations. Weak leadership, bad monitoring of the volcano and a lack of concrete plans for evacuation must take much of the blame. Barry Voigt has a concise, depressing recipe for disaster: 'Armero was caused, purely and simply, by cumulative human error – by misjudgement, indecision and bureaucratic short-sightedness.'[13] The authorities were unwilling to cover the loss an early evacuation would involve, so they waited until it was too late. 'Catastrophe was the calculated risk, and nature cast the die.'[14] Since 1986 Nevado del Ruiz has been the best-monitored volcano in South America.

Lava flows are like bulldozers. They destroy everything in their path. But their slow forward momentum means that people have plenty of time to evacuate even long after the volcano has erupted. Numerous towns situated around volcanoes have suffered the same fate as the Mexican towns of Parícutin and San Juan. Even though the people of San Juan prayed to God for a miracle, the church with its holy crucifix was buried. In Italy as well as in Iceland, the use of the same means has been attempted – with varying success. In AD 253, a lava flow from Etna on its way towards the town of Catania on Sicily was stopped with the aid of a veil. It belonged to St Agathe, who had died a martyr there the previous year. (It was this saint who was rejected in favour of St Francis Borgia as the protector of Lisbon after the disaster of 1755.) The veil had been taken from her corpse and borne in a procession towards the edge of the lava flow, in the hope that it would save the town. A large number of people had gathered, with several of them scourging themselves and others begging for forgiveness in order to prevent the impending destruction. The stream of glowing lava stopped.[15]

Similar processions and the use of St Agathe's veil are well documented right up until the nineteenth century – and the custom still exists on Sicily. An attempt was made to stop a lava flow towards the town of Sant'Alfio in 1971, ending with people placing out their relics and the priest kneeling in front of the lava and asking God to remove the threat.

TRAGEDY IN SEVERAL ACTS

Man's struggle against natural forces has been long and hard on Iceland. The island lies like an open wound in the middle of the volcanic mountain range that rises up from the depths of the Atlantic. This range marks the boundary between the European and the American continental plates that are drifting away from each other at the rate of 1–2 cm a year. And Iceland is gradually being stretched apart, with new land continuously being formed. At several points on Iceland, as along the San Andreas Fault, it is possible to walk across cracks in the ground that define this boundary. One of the fissures is called Laki and it was the scene of the largest volcanic eruptions in history.

Pastor Jón Steingrímsson lived in Siða, only a score or so kilometres from Laki, and he left behind diaries with reflections on the long-lasting disaster and its consequences for Iceland. Could *he* save the church from the destructive lava flows?

'June 8th of 1783, in clear and calm weather, a black haze of sand appeared to the north of the mountains ... That night strong earthquakes and tremors occurred.'[16]

That marked the beginning. Laki was active. The lava streamed out of a 27-kilometre-long fissure, with a score or so cinder cones, or small volcanoes, spewing out volcanic particles and fragments. The eruptions did not stop until eight months later, after having spread a cover of lava over 580 square kilometres of land. For the Icelanders, the 1783 eruption was disastrous. We can follow the course of events from Steingrímsson's account.

10 June: The sky was covered with clouds and the rain was so polluted that people found it difficult to keep their eyes open. The rain smarted when it came in contact with the skin.

13 June: The weather had improved, but the thundering of the volcano had increased and there were frequent earthquakes. Particles and gas streamed out of Laki, polluting the air. 'When it could be seen, the sun appeared as a red ball of fire, the moon was as red as blood, and when rays of their light fell upon the earth it took on the same colour.'[17] The following day, the rain returned and the smell was worse than ever, with many people having problems breathing. Unlike the people of Siða, the migratory birds were able to leave, abandoning their unhatched eggs. Steingrímsson relates that iron rusted and timber lost its colour and turned grey because of the sulphuric acid that fell with the rain. The grass withered. Many people tried to remove the volcanic ash from their fields with rakes, so that the farm animals could survive. Some even washed the grass to get rid of the ash. But it was pointless. The farm animals grew thinner and thinner and gave less and less milk. The horses became so weak that they could not even manage to chew hay, and their heads swelled up. Sheep and cows suffered the same fate and those who tried eating their meat died. Almost 190,000 sheep died in the space of one year. Delays in the supply of grain from the trading centres only made the situation worse, with many people dying of hunger.

15 June: Three farmers climbed to the nearest mountain top to take a look at the eruption. They saw 22 areas with large flames coming up out of the fissure. Many feared for their own farms, including Steingrímsson.

17 June: The first farm was abandoned as a result of the lava flow threat. The farm proved to be inhabited by two men who had moved from the area the previous year in order to be left in peace. As Steingrímsson pointed out, they were very devoted to each other and had 'more than one common interest'. Steingrímsson saw the enforced evacuation as part of God's plan to put a stop to the homosexual pair.

18 June: The lava flows threatened several farms and were as violent as a river during springtime when the snows melt. All the valleys were gradually filled by streams of lava and people were forced to move.

23 June: A pulse of lava streamed down from Laki, with explosions sending new clouds of gas and ash over Iceland. The water went bluish in colour and the acid corroded plants and stones. Steingrímsson was surprised that people had managed to survive at all.

20 July: It was Sunday, and all those capable of walking had gathered in Steingrímsson's church. Lava flows had come close to the church and he feared this might well be the last service to be held there. The church could hardly be seen through the thick mist. Lightning tore the sky, lighting it up inside while the ensuing rolls of thunder sang in the church bells. Even so, everyone felt safe in the church and in God's hands. The service lasted longer than usual, but no one wanted to leave the church. Everyone sat there in deep prayer, resigned, prepared to die. But death was kept at a distance. After the service, they went over the lava flow to see how far it had got. It had completely stopped. Steingrímsson believed it would stay like that 'until the end of the world'. The mood of the people changed. They rejoiced and thanked God for having intervened and saved the church and the houses.

As the year drew to a close, activity on Laki subsided, but the great 'finale' was to take place on 24 December. A huge cloud, almost in the form of a sculpture, rose up. No one had seen anything like it before. It hung motionless in the air, gleaming with blue, orange, red and black. Suddenly, just before sunset, it disappeared. Many people took this as a sign that new problems were in the offing. The volcanic eruption stopped in February, but impure water was still a problem and clouds of gas continued to rise from Laki. Earthquakes were also frequent. At the same time, the famine grew worse. Some people began to eat hay while others sought for old fish bones, boiled them until they were clean and crushed them in a little milk before drinking them. A third of the population in Steingrímsson's parish perished.

Throughout Iceland, people were made painfully aware of their bodies' reactions to the long-lasting eruption: 'Their bodies became bloated, the insides of their mouths and their gums swelled and cracked, causing excruciating pains and toothaches.'[18] The 1780s was a grim decade for Icelanders. People both young and old lost their hair, with diarrhoea, intestinal worms and growths on the ribs, hands, neck and feet making their daily lives intolerable. Finally, the toxic effects were very marked: 'The inner functions and organs were affected by feebleness, shortness of breath, rapid heartbeat, excessive urination and lack of control of those parts.'[19] Steingrímsson realized what was happening. In his opinion, the sufferings and deaths were due to the air pollution and the poisoning of water and farm animals.

'When God acts through nature it is never without purpose.'[20] Despite the fact that Steingrímsson believe that God was behind the volcanic eruption, he never expresses any doubt that God is good: 'May He be eternally praised and honoured for His harshness and His gentleness.'[21] It could have been worse, and Steingrímsson also saw a number of signs that God was trying to lessen the consequences of the eruption. There were, for example, marked changes among those who prior to the eruption had been rich but who after the disaster had lost everything they owned. God's trial of them had actually managed to tame them, in his opinion, 'with the result that they became happier, more humble and more patient the poorer and more impotent they became'.[22]

10,000 Icelanders – or about a quarter of the population – died in the years that followed. We now know that both sulphuric acid and fluorine in the drinking water were part of this large-scale poisoning of Iceland. Archaeological excavations in recent years have confirmed Steingrímsson's accounts, with finds of skeletal deformities – which are a typical sign of fluorine poisoning.[23]

The eruption respected no national boundaries. The wind bore the volcanic ash and the toxic gases to Europe, causing extreme weather and failed harvests. The summer of 1783 was abnormally hot and a dry mist sank over Europe, from Oslo in the north to Casablanca in the south.[24] In Bergen in west Norway the air stank of sulphur, and both plants and farm animals died. On 25 June, seventeen days after the start of the eruption, the mist reached Moscow. In England the average temperature was almost three degrees above normal and the number of deaths doubled in August and September. Many people believed that the mist and the deaths were a sign that the Day of Judgment was at hand. After the heat came the cold. Both the summers and winters that followed were extremely cold. A total of 45,000 Europeans may have died as a result of the volcanic eruption, although the full extent of the eruption has yet to be completely mapped – even more than 200 years after the event.[25]

If Steingrímsson was one of the first to understand the toxic effects on humans and animals resulting from volcanic eruptions, the American Benjamin Franklin was the first to see the connection between climatic changes and volcanic eruptions. He was the American ambassador in Paris and following weather conditions closely. 'The fog was of permanent nature; it was dry and the rays of the sun seemed to have little effect towards dissipating it, as they easily do to a moist fog.'[26] It was not until the 1980s that studies of the connection between volcanic eruptions and climatic changes really accelerated. Today, we know that both ash particles and small drops of sulphuric acid – aerosols – can change the global climate for several years if they reach the stratosphere. The best-known example of this is the eruption from the Tambora volcano in Indonesia in 1815, which led to the 'year without summer' of 1816. The harvests failed and the last major food crisis of the Western world resulted. Malnutrition, epidemics and high food prices, combined with unrest after the Napoleonic wars, led to looting, street riots and violence. Many sought for an answer in religion as to why they had been singled out, the crisis leading to a resurgence of prophecies of doom.[27]

In 1991 the eruption from the Pinatubo volcano in the Philippines caused global cooling that lasted for three years.

MUMMY, WILL WE BURN UP?

The story of man's struggle to stop lava flows does not end with Parícutin or Steingrímsson's prayers. In Iceland in the 1970s prayer was replaced by water in an attempt to save Heimaey from lava flows.

Vestmannaeyjar is a small group of volcanic islands in the southeast of Iceland. The island of Surtsey emerged from the sea in the same area in 1963, as a 'greeting' from the giant Surt. In January 1973 it was Heimaey's turn. It began with a series of earthquakes and the opening up of a 1,600-metre-long fissure past the 5,000–6,000 year-old volcano Helgafell. The continental plates were on the move again. The eruption quickly concentrated in one place, with the formation of a cinder cone – Eldfell – which reached a height of a hundred metres in two days. Ash and volcanic particles were flung into the air and carried by the wind the few hundred metres to the town. Evacuation started immediately with the aid of the fishing fleet, with most of the town's 5,300 inhabitants being transported to the mainland. The evacuation plan went excellently. It was actually only one day old. But Heimaey was quickly transformed into a ghost town surrounded by flames. An observer noted that 'walking around in the streets of the town that first evening was like being reminded of stories of ghost ships that were found drifting on the ocean, with no explanation as to why their crews had deserted them'[28]

The first week was spent trying to salvage the houses from the heavy fall of ash. In February 1974 the eruption of ash died down, the lava flows taking over. The size of the island increased by 20 per cent. The harbour was in danger of being engulfed in lava. Heimaey and the harbour were extremely important for Iceland, as the fleet there was responsible for about 20 per cent of the country's fishing exports. Volunteers set about pumping sea water onto the lava to get it to cool and solidify. In March the lava flowed towards the eastern parts of the town, so more pumping equipment was fetched by boat in an attempt to save the houses and the harbour. In a large-scale attempt to stop the lava, thousands of litres of sea water were sprayed onto the glowing front every second. At one point there was a maximum of 75 people working round the clock to cool the lava.[29] It was the first successful attempt to prevent destruction caused by a volcano. Pastor Steingrímsson would probably have been deeply impressed had he seen all this, but certainly a little ill at ease that the people of Heimaey were trying to defy God's will. He would not, at any rate, have liked the scene that took place a few days after the eruption. A young photo model from a foreign fashion magazine posed for the photographer at the cemetery. She was dressed in the latest fashions, with suitcases full of clothes lying around. The active volcano was used as a backdrop for the photographs.

During the volcanic eruption and the evacuation, the Icelanders displayed drive and flexibility regarding the natural forces. The geographer David Chester has pointed out that their response is typical of modern post-industrial societies. Traditional pre-industrial societies tend to have responses that are co-ordinated by small groups – they emphasize harmony with nature rather than control, call for small amounts of capital only and are extremely flexible. Modern industrial societies tend to have responses that are the opposite of this. Responses that are typical of post-industrial societies, such as Iceland in 1973, combined the traditional and the industrial responses to forge a more tailored and flexible system for dealing with the needs of a specific society.[30] A criticism of this analysis is based on the fact that it is somewhat deterministic – that it functions as a kind of general formula for how societies react to crises.[31]

The 1973 eruption has acquired a key position in people's awareness – and in the society at large. A social anthropologist who has studied the culture of Heimaey claims that the event has become an important part of people's identity. The natural disaster has become institutionalized and functions as a source of new rituals and traditions.[32]

The anniversary of the death of the volcano is celebrated on the first weekend of June every year. Parades are organized, with young people dressing up as small volcanoes. Reminiscences of the eruption are told in the evenings. The disaster has become an important part of the tourism industry and been incorporated into the popular culture. A film about the eruption is shown and people can join in making miniature volcanoes. The popularization and dehazardization of the eruption are events that ensure the eruption a place in the collective memory. The eruption *will* be remembered. The physical reality is still evident – smoke continues to come from the volcano. This attracts tourists and thus the eruption has also become important for the economy.

The eruption from Heimaey lasted for five months, burying about 360 houses. The only physical injury incurred was a broken finger. But what did the people feel about the eruption and the risk of living on a volcano? In a questionnaire carried out among those evacuated after the eruption, 10 per cent said that they did not want to return to Heimaey. It was the elderly who found it most difficult to cope with the eruption and who wanted to remain on the mainland.[33] Generally speaking, it is normally old people who are hit hardest by natural disasters and the swift changes they cause. Perhaps they do not have enough strength or means to find new solutions and re-establish themselves. Young people are vital for the re-establishing of disaster-stricken societies.

LIVING WITH RISK

Can the volcano erupt again? Both those on the mainland and Icelandic geologists believe that Heimaey is still a risk area. But most islanders refuse to accept that a similar disaster can take place again. It is as if they have made a choice in order to be able to survive on the island at all. All doubt must be removed. So when the youngsters ask 'Mummy, will we burn up?', their parents reply on the basis of their personal choice that no danger exists. The reactions of a woman from the island are a good example:

No, a volcanic eruption out here on the island . . .? That's something that . . . I don't think will happen again. Not again, not again. Not in my lifetime. I am *absolutely certain about that!*[34]

People claim that the situation is safe and have to defend that view of the world. From that point of view, the apparent absence of fear is a social and psychological phenomenon, as it was in the fjord areas of west Norway that were exposed to slides. This contrasts, for example, with the view of nature in California, where it is the *fear* that is the social construct.[35] The ambivalent attitudes to the volcano and possible future eruptions is also expressed in the attitudes of the inhabitants to nature and the volcano itself – Eldfell. The inhabitants are divided in how they view the new landscape that suddenly appeared in 1973. Newcomers think that the volcano is impressive, while those who knew the landscape prior to the disaster feel that it has become ugly after the eruption.[36] The volcano destroyed the green meadows, leaving behind a barren landscape.

A 64-year-old man from Heimaey explains his choice: 'I am well aware of the risk but I accept it as it is! Everywhere on Iceland there is a danger of natural disasters, so where should one go? You can't be on the run all your life.' Only five people who took part in the questionnaire in 2003 felt it was highly likely that there would be new eruptions from the volcano.[37] All of them were men. The most widespread reaction to the thought of new eruptions was despondency.

Fatalism is common on Heimaey and it found expression in the way many people confirmed that the future is determined by God's will. The Pentecostal Movement has been strong on the island for a long time, with three times as many members compared to the average percentage on Iceland. In the 1930s, a sect also came into being on the island after a woman had had a vision that God would punish them with an eruption if their sins continued. A man from Heimaey stated that the sect 'spread prophecies such as "Be prepared – God is angry!", so that when the eruption occurred there were many people who were not surprised.'[38]

A classic study from the Mount Saint Helens eruption in the USA in 1980 shows what can be at stake after a volcanic eruption – or any natural disaster for that matter: people's mental health. Initially, a natural disaster can lead to post-traumatic stress disorder because it becomes difficult to known how to deal with the risk. In the small city of Othello in Washington, much time was to pass before things got back to normal. The ash streamed from the volcano on 18 May 1980 and was borne on the wind to Othello. Via information from clinics, hospital and police, psychologists ascertained that there were major differences in the way people behaved after the eruption. The most dramatic outcome of people's problems was a 47 per cent increase in domestic violence. The consumption of alcohol shifted to the home because conditions for driving a car were difficult. At the same time, there was a large reduction in reported abuse of children – something that indicated a strengthened sense of family. The crisis telephone service registered an 80 per cent increase, and psychologists also had a lot more to do. People's problems were stress reactions linked to a fear of the unknown, the volcano and the fall of ash.[39]

NOT QUITE AS SUPER

In accounts of volcanic eruptions and how they have influenced us, one cannot omit the so-called 'supervolcanoes'. These volcanoes have caused enormous disasters and in recent years have become a favourite theme for the disaster-hungry media industry. Supervolcanoes differ from other volcanoes in more ways than just the intensity of the eruptions. In terms of area they are larger than normal volcanoes and often cover large areas without any distinct cone-shaped mountain. Furthermore, eruptions are extremely rare. The best-known supervolcano is Yellowstone in USA, where the last eruption took place some 600,000 years ago. The most interesting one, historically speaking, lies on the other hand in Indonesia and is called Toba. A volcanic eruption from it about 71,000 years ago is one of the most powerful we know of, with new knowledge on its effect on humans having come to light in recent years. The eruption was so huge that it caused a cruel winter to hold the earth in an iron grip for several years. This knowledge is based on data from ice cores in Greenland, which have shown high concentrations of sulphur in strata that corresponds to a period of six years. The volcanic winter was created by particles and aerosols that were flung up into the stratosphere. At high latitudes, the summers may have been twelve degrees colder than usual. India was covered by a layer of volcanic ash that was up to six metres thick. Toba may have spewn out as much as 800 cubic kilometres of lava and particles. That is about 40 times as much as the Santa María eruption in 1902. The anthropologist Stephen Ambrose is among those who believe that the eruption almost wiped out the human race.

One model of human development is that mankind spread out from a core area about 100,000 years ago. Based on a modelling of genetic variations in present-day man, researchers have claimed that the population of the earth was reduced to some few thousand individuals at the same time as the eruption occurred. The number of survivors is controversial, with anything from 50 to 10,000 fertile women being advanced. Spread out over the land masses of the world, they must have had plenty of room. About 20,000 years after the eruption, mankind began to flourish in genetically distinct groups. Evolution takes place faster in isolated population groups, so without the volcanic eruption our genetic variation would probably have been less. Everyone living today would probably have been more like present-day Africans. The so-called genetic 'bottleneck' may thus be a direct consequence of the eruption and the ensuring volcanic winter.[40] For evolution and the development of new characteristics in species, dramatic upheavals are more important than long-lasting stable periods. Natural disasters can be important 'engines' of evolution.

If one is prepared to go a long way back in thinking about volcanic eruptions, then 252 million years is a key figure. Most of life on earth died out at the same time as a gigantic volcanic province was formed in today's Siberia in Russia. Lava ran out of fissures, covering an area corresponding to ten times the size of Norway, with a six-kilometre-thick layer of lava and volcanic material. Most sea-life disappeared (95 per cent), as did over a third of life on land.[41] What the animals died of is still one of the great unanswered questions of the geo-sciences. But if we think about what we have learned from the volcanic eruptions of Toba and Laki, it is perhaps possible to understand a bit more. Endless winters, little sunlight, enormous quantities of toxic gases, polluted water and the destruction of land areas were all factors. Those that could not flee or adapt vanished for ever. This opened up new niches for others. No one knows what life on earth would have been like today without the enormous volcanic eruptions.

> The Mountain crept up ever nearer, until, if they lifted their
> heavy heads, it filled all their sight, looming vast before them: a
> huge mass of ash and slag and burnt stone, out of which a sheer-
> sided cone was raised to the clouds . . . [Sam] raised his eyes with
> difficulty to the dark slopes of Mount Doom towering above him,
> and then pitifully he began to crawl forwards on his hands.

The hobbits Frodo and Sam have to throw the ring into the glowing volcano known as 'Mount Doom' in order to save the world.
The world of good is being threatened with destruction controlled
by a volcanic flaming inferno. As in Norse mythology, the natural
forces of J.R.R. Tolkien's mythological world are controlled by the
forces of the dark. The volcanoes in *The Lord of the Rings* symbolize
destruction and dark powers. The destruction of the ring finally
leads to the victory of good over evil, while the volcano pours out
lava and is shaken by earthquakes during its death agony. 'Fire
belched from its riven summit. The skies burst into thunder seared
with lightning. Down like lashing whips fell a torrent of black rain.'
Tolkien was fascinated by man's struggle against nature and natural forces. In *The Lord of the Rings* he also played on the generally
negative attitude of the Western world to volcanoes.

It is particularly in Western culture that volcanoes are connected to something destructive. In other parts of the world
mountains and volcanoes can have a completely different symbol-
ism than in Western culture. In India there are a number of holy
mountains. Within Hinduism, it is common for mountains to be
seen as divinities or as dwelling-places for gods. Himalaya, for
example, is the father of the goddess Parvati, whose name literally means 'daughter of the mountain'. Tirumada in southern India
is another holy mountain, with a number of small temples locat-
ed on the path to its peak. The climbing of the important pilgrim's
route is meant to give a feeling of ascending towards the divine.[42]
In Japan there is similarly a number of volcanoes that are worshipped. The best example is Mount Fuji.

Everyone thought that Mt Fuji was extinct, right up until a
series of earthquakes at a depth of over ten kilometres beneath the
volcano was measured in October 2000. The last eruption was in

1707 and it gave geologists who have studied the volcanic deposits plenty to think about. For it has emerged that explosive eruptions and flows of hot ash were normal, despite the fact that Mt Fuji belongs to a type of volcano that usually only has peaceful eruptions. The mountain can be seen from the centre of Tokyo in clear weather, so eruptions of ash from the volcano can have dramatic consequences. The distance to Tokyo is only 100 kilometres.[43] Dangerous or not, Mt Fuji is Japan's number one national symbol – and is moreover a holy mountain. In Japan, mountains are often holy, some of them so sacred that it is considered blasphemous to climb them, even for priests. In several Japanese religions mountains occupy a central place, both within cosmology, burials and rituals – Buddhism and Shintoism being the most important. Within Shintoism the traditional animistic belief was that all natural phenomena were dwelling places for spirits – or *kami*. There was mutual interaction between *kami* and humans. Gradually, Buddhists also became interested in the mountains as metaphors of spiritual enlightenment. About a thousand years ago Buddhist and Shintoist temples existed side by side at the foot of a number of mountains in Japan. At the same time, a new hybrid religion emerged, called *shugendo*. Its practitioners were ascetics that lived a secluded life in the mountains, one of their aims being learning mastery of magic forces. Towards the end of the nineteenth century, *shugendo* became a popular folk religion. Mt Fuji, with its 1,300 temples, has been a popular pilgrimage destination since the fifteenth century, with 400,000 people climbing its sides every year. The ascent is divided into three stages of unequal length, with temples scattered over the slopes of the mountain.[44] At the summit, the crater, there are no longer any small temples. Nor any buildings. The entire mountain is a *kami*.

Culture and religion greatly influence how people understand and interpret danger. This is particularly true of the most dramatic of natural phenomena – the volcanic eruption. When lava and ash pour out and the sky is rent by lightning, the fascination is almost unreal. It is easy to understand that people have made use of religious and metaphysical explanations as to why they exist and what attitude ought to be adopted to them.

7

The Politics of Disasters

'He wished to behave like all those others round him
who believed, or made-believe, that plague can come
and go without changing anything in men's hearts.'
Albert Camus, *The Plague*[1]

Western Europe enjoyed a warm and pleasant climate during the
1730s. Some of the average summer temperatures that decade
almost matched the record warmth of the 1990s. This was not meant
to last. For some reason, the climate changed dramatically during
the autumn of 1739. In Britain crop yields were low due to cold and
wet weather. The winter of 1739/40 became one of the coldest ever
recorded in Europe. A record low temperature of -18°C hit London
and the Thames was frozen for seven weeks in January and
February 1740. The harsh winter was not restricted to Britain, but
affected most of the countries in Western Europe. The arrival of
spring did not make things better, and the cold weather persisted
throughout the summer. In Denmark the clergyman Erik
Pontoppidan wondered why nature had lost some of its vitality
that year: 'it was as if the rays, warmth and reviving power of the
sun had decreased, and the earth's crops stood there languishing,
without being able to ripen.' The average temperature in central
England in 1740 was as low as 6.8°C – the lowest since records
began in 1659. It was also the driest year for 282 years.[2]

The consequences of the cold and dry summer were felt during the autumn. Large parts of Europe had been hit by crop failure.
The corn was nowhere near ripening, although it was late autumn.
In many regions, especially in Scandinavia, the yield from the harvest was absolutely nothing. The price of grain rocketed as a result
and the common people went hungry. Grain prices rose by more
than 70 per cent in Denmark and the Low Countries.[3] Food riots
and granary robberies in France, the Low Countries and England
shows that people were desperate.[4] Hundreds of thousands were
fighting for survival. The additional consequences of the food

shortage became important that very same year. Many small farmers and poor people had to abandon their farms and moved to the towns, where they looked for a job or had to beg. If worsened climatic conditions provided the initial trigger, epidemics spread by aid-seeking people on the move ended in a mortality crisis. Spotted fever, dysentery and smallpox were common causes of death, and the first epidemics hit early in 1740. There was a dramatic 21 per cent average increase in the number of deaths in Europe that year.[5] In London, death rates rose by 53.1 per cent from January to June 1740 compared to the previous winter and spring.[6] The countries hardest hit by epidemics were Norway, Finland, Sweden, Scotland, Ireland, England, France, the Low Countries, Italy and Switzerland.[7] Germany and Austria also suffered from rising grain prices, but managed to avoid a mortality crisis.

Everything got worse in 1741 following another hard winter and spring. Many people had frozen to death. As the summer of 1741 came to an end, farmers had to harvest the corn while it was still unripe. The price of wheat, oats and rye increased further. Many people used up their last resources in an attempt to procure something to eat, and people paid with 'Sunday clothes, even shirt buttons, hooks & eyes and the like'. Their diet was eked out by bark, grass and earth.[8] By the time the climate normalized in 1743 and 1744 and the crops were harvested, Europe had been through one of the worst crises since the dramatic late 1600s. It is not known how many people lost their lives.

The handling of the crisis in the 1740s varied from one country to the next. In Sweden, Finland, France, Ireland and Denmark the harvest reduction was critical. However, the historian John D. Post has shown that it was social and political factors that converted drought and cold weather into famine, epidemics, and death: insufficient control of food prices and grain reserves, poorly developed infrastructure and welfare systems, and a low standard of personal hygiene. In England it was an export ban on grain that eventually saved lives, the same being introduced in Baltic, Russian and German-speaking areas. But it was first and foremost in Scandinavia and Ireland that mortality rose dramatically. As many as 400,000 people may have died in Ireland.[9] Short-sighted emergency planning and a lack of welfare systems helped to create the crisis in Ireland.[10] Also, the dependence upon potatoes as main food source made crop failure more critical. In Norway, which had the highest mortality rate of all European countries (81 per cent when compared to the 1735–9 average), the delayed action taken by the ruling Danish authorities worsened the crisis.[11] In Finland, a war between Russia and Sweden contributed to food shortage and increased mortality.

THE LITTLE ICE AGE?

What caused the climate to change abruptly between 1739 and 1740? What were the natural events leading up to the unusual temperature anomaly? It has been argued that the reason for the cold weather was a sudden change in the North Atlantic weather pattern, or the North Atlantic Oscillation (NAO). High pressure systems over Iceland and Greenland resulted in cold weather in much of Western Europe and freezing winters. This opened up atmospheric pathways for cold winds from Siberia and the Arctic to flood Europe.[12] It has also been suggested that the climate change was a part of the general cold climate of the 'Little Ice Age'. However, we are still left with the question of what caused the abrupt change from the warm 1730s to the record-breakingly cold 1740. A powerful volcanic eruption in August 1739 from Tarumai in Japan released huge volumes of ash and sulphur into the atmosphere.[13] But was it enough to trigger changes in the North Atlantic region? Probably not. The 1740 event remains enigmatic and poorly understood.[14]

Regardless of what triggered the change, the crisis of the 1740s provides an insight into how vulnerability to climatic changes increases when food is in short supply and the welfare systems are dysfunctional. Hunger and epidemics are often caused or worsened by combinations of extreme natural phenomena (abnormal frost periods, floods and drought) and social and political factors. No famine takes the lives of the rich. It is a problem of poverty. Thus the resulting famine is just as much a manmade phenomenon as a natural disaster. One of the characteristics of famine-triggers compared to geophysical hazards is that they develop over a long period of time and can therefore be detected – and, in theory, prevented. It took however more than a hundred years until the welfare systems in Europe had developed sufficiently in order to prevent widespread famine. In fact, Western Europe experienced another major mortality crisis in the early 1770s, again caused by abnormal weather, failed harvests, and the spread of epidemic diseases.[15] How badly could things have gone wrong in the 1740s without the strong governmental control on food distribution in England, France and the German-language states? Would merchants have sold grain to the highest bidder, thereby keeping more people away from the existing food reserves? The winter of 1739 could have resulted in a 10 to 20 per cent population loss in Western Europe, as it did in Ireland and Norway.

SCORCHING HEAT

After the disastrous famine in Ireland in 1845–50, widespread famine left Europe, never to return. However, the improved food security and health conditions in Europe towards the end of the nineteenth century were not mirrored in other parts of the world. One of the worst disasters ever was triggered by drought, failed harvests and the imperialistic iron grip of the West on global food production.

It was October 1876, and both the British colonial power and the majority of India's millions of inhabitants were hoping that the monsoon rains would save that year's crops of rice and wheat. The previous years had been good – the surplus had been exported to England and in the streets of London people were eating bread baked with Indian grain. But the monsoon, with its much needed rain, did not come. This resulted in a number of disturbing events towards the end of the year. People left the countryside, where they had eaten rats in order to survive while waiting for rain and the harvest. At the same time, granaries in towns were full. British investors and fortune-hunters had gambled all they owned on buying grain in the hope of making a fortune. Transportation of food to the affected areas ought not really to have posed any problem. Considerable sums of money had been used to build railways in India, one of the official reasons for doing so being precisely that it had to be possible to transport food to the countryside in the event of a food shortage. Instead, the railways were used to transport grain *out* of the very areas that needed it. Through the use of telegraphy, the prices of grain and rice were kept equally high throughout the country. Prices shot up dramatically, so that grain was soon effectively out of reach of the increasingly hungry and poor part of the Indian population.[16]

As the year progressed, conditions got drastically worse, the lack of food spreading northwest from Madras towards Bombay. Thousands of people flocked towards the towns, but were stopped by the police and denied entry. People died outside the supplies of food. The situations could have been prevented by the British Empire's man in India, Lord Lytton. But he refused to interfere in the free market that controlled the grain prices. He himself lived in the lap of luxury in Delhi and was preparing what was to be the world's largest and most expensive meal ever. The occasion for this was that Queen Victoria was to be proclaimed Empress of India. No less than 68,000 guests enjoyed the most delicious of dishes for a whole week while 100,000 people were dying of starvation in the south of the country. Lytton was to be remembered by Indians as 'the British Nero'.

During 1877 the sufferings of the poor increased. The British were still unwilling to intervene in grain prices, exports to England actually increasing to twice what they had been in 1876. Revolts spread in parts of the country, with many desperate attacks against the high prices. Good harvests from before 1876 had been exported; furthermore, agriculture in parts of India had switched to cotton production instead of corn. In the province of Deccan people now felt the consequences of being part of the global market. In some places people were so desperate that they sold their children for just a little food. Stories of cannibalism are legion.

The summer of 1877 was a scorcher, but rice harvests from Burma and Bengal were good. Rice was exported at the same time as India now officially had 36 million people suffering from extreme hunger. Furthermore, a cyclone had claimed the lives of 150,000 people in Bengal and given rise to a violent cholera epidemic that affected practically everyone. It became a pandemic. In Bombay attempts were made to prevent leakages to the press about the enormous death toll. Corn revolts were everyday occurrences and prisons were full to the brim. Those worst affected in the highly stratified Indian society were the lower castes and the untouchables. Women and children caught stealing food from the fields or granaries risked torture, having their noses cut off or being killed. The only people who seemed to get on all right were

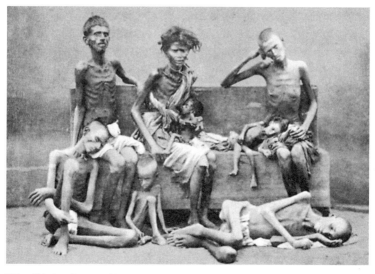

Why did they have to die? Between six and ten million people in India suffered the same fate as these eight people in the 1876–9 period – they starved to death. Photographed by Willoughby Hooper.

the wealthy. In October the drought came to an end, but downpours led to excellent conditions for mosquitoes. Malaria claimed several hundred thousand of lives. There was no end to the tragedies. Those who survived the years of crisis until 1879 now, on top of everything, received visits from armed tax collectors, who wanted debts that had now soared to impossible heights to be repaid.

In *Late Victorian Holocausts* the radical historian Mike Davis has described the ideology underlying the attitude of Lytton and the British to the famine disaster. Aid to the Indian population was not given priority. Malthusian principles regarding population growth (food production cannot keep up with the increasing population figures) were mixed with social Darwinism (the fittest survive and the weakest will die in any case) in order to justify the lack of action. It was 'Nature itself' and the 'far too high a reproduction rate' of the Indians that led to so many of them dying. Comprehensive aid would only lead to even more needing help later on. This point of view was repeated by a British commission set up to investigate the causes of the disaster – and it was endorsed by the British parliament.

Despite the fact that people were dying of hunger throughout large parts of India, the empire was never without food. Many were starving to death while wagons loaded with grain rolled past. A number of camps for the hungry were set up, but these only made the situation worse. Food rations were carefully calculated by the British authorities, but they were too small to live off. Furthermore, everyone had to work hard for their food. This led to emaciated Indians dying in great numbers from starvation and epidemics, even though they were actually supposedly under British protection. Lytton, who had previously shown nothing but contempt for those suffering in India, visited a camp for the first time in August 1877. Did this open his eyes to the human tragedy? The answer came in a letter to his wife after the visitation. He stated, on the contrary, that life in the camp was like a party: 'The people in them do no work of any kind, are bursting with fat, and naturally enjoy themselves thoroughly.'[17] Lytton was not being ironic. He had a completely racist view of the Indians. Over the next three years the Empire would have between six and ten million lives on its conscience.

The public was deceived until 1878. It was then that the critic Robert Knight visited India, subsequently describing the Empire's famine policy as 'murder'. After this, criticism hardened. But the report from the famine commission completely exonerated the authorities, even though the opposite could be read in the newspapers. 'Why blame poor Nature,' as the Indian economist and nationalist Dadabhai Naoroji expressed it, 'when the fault lies at your own door?'[18]

THE FORMATION OF THE THIRD WORLD

Eighteenth-century India and China had far superior routines for dealing with drought, flooding and famine than European countries. In China the emperor personally followed developments in food prices from month to month. Furthermore, living standards in Asia at that time were at least as high as in Europe. In China engineers who were behind successful flood-prevention measures were acclaimed as heroes and honoured with their own temples. Large-scale famines and epidemics were avoided by food being effectively distributed to those in need. Aid was given high priority so that natural disasters did not become too extensive. The tropical areas of the earth dominated the production of commodities. While Europe only accounted for 23 per cent of the world's production in 1750, the remaining 77 per cent came from the tropics.

Why did famine appear in Asia in the late nineteenth century when it at the same time disappeared from Europe? The reason is not that India and China were weak in resources or poor, Mike Davis claims, but that the countries in the tropics had both their labour and products subjected to a world economy controlled from London. Famine was not caused by these peoples existing outside the modern world, but, on the contrary, because they were actually 'forcibly incorporated into its economic and political structures'.[19] The tropical countries were reduced to primary produce suppliers to the industrialized 'North'. During the period of most rapid expansion in Europe, between 1750 and 1790, India's economy stagnated completely. The high prices of food transformed drought into famine. In this process, 'The Third World' was born, conceived by extremes of weather, imperialism and the transition to a capitalist world economy. The inequality between nations was a modern invention on a par with electric light and the machine gun, Davis points out.

REPETITIONS

The famine of 1876–9 did not only affect India – it was global. In China, Egypt, the Philippines, Korea, Brazil and southern Africa the period of crisis was preceded by extensive drought, floods and crop failure. The number of deaths in India and China probably exceeded 20 million. Even so, this was only the beginning of the problems. New periods of ruined harvests caused by drought and flooding affected the southern hemisphere in 1889–91 and 1896–1902. In India, Korea, China, Ethiopia, the Sudan and Russia people suffered from a lack of food. In India the British Empire repeatedly avoided taking measures to reduce the country's vulnerability to the disasters. Starvation, mixed with epidemics of cholera, dysentery, malaria, bubonic plague and smallpox, claimed new victims. During the three periods of drought, the earth lost between 30 and 60 million inhabitants.[20]

In China civil war had emptied the provincial coffers. When the harvests failed, the granaries were empty, and a bad infrastructure and transportation network meant that it was difficult to get emergency supplies out to the provinces. Ninety million people were affected. 'Children abandoned by their parents', it was said, were 'taken to secret locations, killed and consumed.'[21] Entire villages migrated southwards to beg. The phenomenon was so widespread that it was given its own name: *t'ao-fang*. The hardest hit area, Shanxi, did not reach its original population figure until 1953.

During the drought and famine that arose in India in 1896, Lord Lytton's attitude towards the Indians was repeated, this time by Lord Elgin. The lesson to be learned from the earlier crisis had not stuck, and an efficient distribution of grain at market price ensured that prices were equally high even in areas that had not been affected by drought. This meant that for most people food was beyond reach. Instead of regulating prices and lowering exports, Elgin introduced the poorhouse. The food served there was dry flour with salt. The poorhouses were deeply hated. As if this was not enough, bubonic plague arrived in Bombay in summer 1896, decimating the lower castes. The British authorities did all they could to prevent the disease spreading, even though this meant treating people brutally. They burned down houses and disinfected and forcibly evacuated all suspected carriers to their own camps. Only a few people came out of them alive, many Indians being convinced that they had been set up in order to do away with them and that their bodily oils were to be used for producing cosmetics in Europe. Public criticism of the 'disinfection campaign' was forbidden. It was, however, not particularly effective. No fewer than 6.5 million people died of famine and epidemics in 1898. The plague was difficult to get rid of and it took the lives of a further 8 million people in India up to 1914.

Racist attitudes to people affected by natural disasters and crises are nothing new and they have on a number of occasions contributed to the legitimization of an occupation: 'They are unable to take care of themselves.' In former centuries, the belief that God was behind natural disasters was also widespread among the higher echelons of society, as we have seen in earlier chapters.

When people in occupied countries in Latin America, Asia and Africa were affected by natural disasters (earthquakes, volcanic eruptions, drought and floods) and were apparently punished by God, this had to mean that they were sinners that ought to be subject to and educated by Europeans. One of the reasons why poorhouses were established was that Indians were thought to be skivers and beggars by nature.

Eighteenth-century India must have seemed almost to be a heaven on earth. Europeans there greatly appreciated the abundance of food, the good climate and a rich literary and scientific tradition. This impression changed drastically during the latter half of the century. After a drought and a famine in 1770, the attitude of the British towards the Indians altered completely. They were now seen as being feeble, weak and lackadaisical. No fewer than ten million people died as a result of the food shortage (high grain prices) and epidemics. Historians have claimed that this contributed to the British legitimizing their presence and control in India. The West considered itself superior to Asia. While the Asians were passive and lazy by nature and without control of natural forces, people in Europe claimed to be in full control, even when chaos raged. 'Let us now represent to ourselves', Abbé Raynal wrote in 1777, 'any part of Europe afflicted by a similar calamity. What disorder! What fury! What atrocious acts! What crimes would ensue!'[22] In other words, passivity was not a highly appreciated quality in eighteenth-century Europe. Chaos and looting were positive reactions compared to the 'lackadaisical' attitude of the Indians. So it did not help to accuse the East India Company of contributing to the famine through the corn monopoly. A gulf had opened up between Europeans and Asians that later was to develop into an abyss.[23]

OCCUPATION AND PRAYER

Colonial powers made effective use of the enormous crises of the nineteenth century. Countries ravaged by famine and epidemics were easy prey. European powers, Japan and the USA systematically acquired new colonies, new land and new resources while the peoples were down on their knees. In the Philippines hunger and starvation were part of the Americans' military strategy. They hampered supplies, closed ports, killed farm animals and destroyed rice stores. In other parts of the world Europeans displayed a similar degree of inhumanity. 'Europeans track famine like a sky full of vultures', a missionary was told by an African.[24] The statement was very apt. Famines in the late nineteenth century were like a godsend for the aggressive British imperialists, Mike Davis points out.

Missionary activity was another, more indirect way of gaining influence. The famine in China in 1876–8 was seen as a 'heaven-sent opportunity to spread the gospel'. Pressure was applied to evangelize China as quickly as possible and to exploit 'the wonderful opportunity' that the famine had created. The British consul believed that the work of the missionaries would help to open up China more than tens of wars would have done. The conversion of China to Christianity did not, however, go as easily as people had anticipated and the results were poor, with only few converting. Many Chinese believed that the missionaries and Christianity had disturbed the balance of the earth so that 'the dragon of the earth' had woken up and caused drought and flooding. Rain failed to come because the churches had 'sealed the heavens'. Farmers were told by Buddhist priests that the drought would not disappear as long as the Christians interfered with Chinese traditions. Disaster evangelism is by no means a phenomenon that only belongs to the past. After the tsunami disaster of southern Asia in 2004 attempts were made to entice Buddhists on Sri Lanka over to Christianity with the aid of expensive presents and money.[25] Similar stories of missionaries having exploited the 'opening' provided by the disaster have also come from Indonesia.

In Zululand, in Africa, the king thought that there had to be a connection between the crisis and the British invasion: 'I feel that the British chiefs have stopped the rain and that the country is in the process of being ruined.' In Ethiopia people had no explanation as to why they had been repeatedly hit by disasters. A lack of piety was suggested as an explanation by Emperor Melenik, who exhorted his people to pray. In Moslem, Christian and other religious communities people waited for the year 1900 with dread. Throughout the world, natural disasters and famine were seen as symptoms of a gigantic crisis. They became symbols of imperialism.

The food shortage and famine of the late nineteenth century led to desperate reactions in a number of countries. In India radical nationalists rebelled against the British Empire. In China explosive mobilization took place in summer 1900 as part of the so-called Boxer Rebellion in the areas worst affected by the drought. Thousands of supporters of the humble Doomsday prophet Antonio Conselheiro were brutally massacred in the Brazilian town Canudos. The drought and famine in South America helped to mobilize popular Catholic apocalyptic attitudes. The end of the world was nigh. In both eastern and southern Africa a number of revolts were linked to the drought and famine. In many instances they were motivated by Doomsday prophecies. In the Philippines, Brazil, Korea and Vietnam the famine gave rise to popular religious movements, with many people using magic rituals to try and get the rain to return. Adventists joined forces with nationalists in opposing the colonial powers.

EL NIÑO

The periods of drought in Asia at the end of the nineteenth century were caused by fluctuations in the global climate system that are known as El Niño – Southern Oscillation, abbreviated to ENSO. These are caused by changes in the temperature of the Pacific around the equator. Normally, cold water streams from the coast of Peru towards the warm water to the west in the Pacific, where the air rises and streams back towards Peru. Occasionally, the whole system reverses as a result of the Pacific being warmest near Peru. Then the warm sea currents move eastwards while the winds are spread both eastwards and westwards. The accumulation of warm water off Peru has been given the name El Niño, 'the little boy'. Combined with the airstreams and the wind system it forms ENSO, which gives rise to the strongest natural short-term fluctuation in climate that we know of. It was ENSO that triggered the crises between 1876 and 1902.

ENSO has global consequences and can influence other weather systems, also round the Indian Ocean and through South America to the Atlantic. El Niño is frequently followed by an opposing phase known as La Niña, which is characterized by masses of water off Peru that are colder than normal and heavy downpours and floods, whereas El Niño gives rise to drought. During the twentieth century 23 El Niño and 22 La Niña occurrences were recorded. In

1973, 1983 and 1998 the effects were particularly strong, with dramatic consequences for millions of people around the world. In Africa there were droughts both in Sahel and Ethiopia. The most powerful El Niño ever to be recorded was that of 1998. Drought affected Australia and southeast Asia, and almost two million people in Brazil had to have emergency aid in order to stave off famine.[26] Occurrences of ENSO have become both more frequent and more pronounced since the 1970s. A possible cause is global warming.[27]

The combination of extreme weather phenomena, political misrule and a vulnerable population have also led to enormous losses of human life since 1902. The famine in the Ukraine and other parts of the Soviet Union in 1931–3 was triggered by drought and failed harvests. Combined with Stalin's harsh political regime and a strong desire for industrialization, the population was badly hit by the drought. Almost six million people died.[28]

The extent of the disaster has only been surpassed in Mao's China. Mao's reform programme of the 1950s – The Great Leap Forward – was to change China from being a nation of farmers to a nation of iron manufacturers. The reform coincided with the worst drought of the twentieth century. This resulted in food shortages and famine. Combined with governmental concealment and the absence of an opposition party, this resulted in the death of 20–40 million people in 1959–61.[29] The disaster was denied by the authorities, who even publicized news of record harvests.[30] It is difficult to understand that one of the greatest disasters ever to have affected China took place such a short time ago.

SUSTAINABLE DEVELOPMENT

Especially today, drought and floods are the natural disasters that have the greatest scale and the most far-reaching consequences. It is difficult to ensure access to food for those in need when political instability, drought, small food stocks and short-sighted solutions are widespread. In the 1970s and '80s Africa was particularly exposed, the combination of drought, wars and unstable regimes resulting in the loss of over one million lives. A ray of hope is that a number of potential famines have been avoided by the alarm being given early and emergency supplies being sent in.[31] There are examples from Bangladesh and Africa from the 1970s onwards.

As we have seen, instances of famine in both Europe and India were caused by a combination of natural, economic, political and social conditions. Famine always has complex, closely knit causes. There is no single model or theory that can explain everything. There has, though, been a tendency over the last few decades for famine disasters to be increasingly triggered by politically unstable countries that are racked by internal conflicts. One of the most respected researchers within this area, the Nobel Prize winner Amartya Sen, has claimed that neither drought nor ruined harvests inevitably lead to famine. The most important thing, Sen claims, is the weakening of people's possibility of buying food that otherwise is accessible – for those with money. The type of government is also very important. One of his assertions is that democratic countries have never been hit by famines. The main reasons for this are considerable freedom of the press, social security and the presence of a political opposition.[32] This view is supported by the fact that many famine disasters, especially in Africa, have not had anything to do with extreme weather conditions. In other instances, drought can be the stress factor that knocks a fragile system off balance. The geographer Mark Pelling has pointed out that the concept 'environmental disaster' is a more appropriate term than 'natural disaster' since an intimate interaction between nature and society helps to increase vulnerability. Famine and other disasters (slides and floods) that occur frequently in poor countries with overpopulation, deforestation and non-sustainable development can therefore also be called environmental disasters. The theme is still the subject of debate, but in principle we can prevent climate changes and droughts from developing into disasters in the same way that the consequences of earthquakes can be reduced. The future of millions of people is at stake if we do not succeed.

8

Climatic Disasters

'Warming of the climate system is unequivocal.'
Intergovernmental Panel on Climate Change, 2007

Man has influenced the earth's climate for thousands of years. Ever since the agricultural revolution about 8,000 years ago, we have altered the concentration of greenhouse gases in the atmosphere. Farming meant clearing forests and fields, causing emissions of carbon dioxide. Furthermore, cultivated fields retained heat better. Farm animals and rice fields were also sources of methane emissions. The American researcher William Ruddiman has claimed that the Little Ice Age was caused by a decline in agriculture after the Black Death.[1] There were fewer people to cultivate the land, farms lay abandoned and an increase in carbon dioxide was effectively halted when trees and plants conquered the cultural landscape. If this is correct, the interaction between mankind, culture and nature is even more complex than previously assumed.[2]

Until the mid-1970s there was a widespread belief among researchers that the global climate was becoming colder and that we were possibly heading for a new ice age.[3] Twenty thousand years ago northern Europe and northern America were completely covered by ice. The possibility of the ice-caps returning constitutes the 'ultimate' climate threat. The glaciation of the world was the scenario of the 2004 American disaster film *The Day After Tomorrow*. Even though the film was criticized for not being scientifically correct, a new ice age would actually be capable of driving people away from the 'rich north' to the 'poor south'. As knowledge of the earth's climate increased during the 1970s, it appeared that the earth was in fact getting *warmer*, not colder. It was claimed that the cause of the warming could be manmade emissions of gases such as carbon dioxide and methane, which affect the earth's heat balance. The observing station Mauna Loa on Hawaii had documented an increase of carbon dioxide in the atmosphere of over 4 per cent between 1960 and 1975. But the consequences of this were uncertain, and few claimed to have all the answers. A few people still warned of the advent of a new ice age. Others, though, claimed that global warming would have dramatic consequences. While researchers were working on trying to understand the effects of the emission of greenhouse gases, the issue became politicized. Ought our dependence on fossil fuels to be reduced? How much money ought to be spent on preventing global warming? The debate was heated, especially in the USA. Conservative politicians characterized assertions of the possible negative consequences of the emission of greenhouse gases as left-wing propaganda. The National Academy of Sciences in the USA warned in 1977 that the earth's temperature was rising and that the consequences of this in the future, that is, in a hundred years' time, could be disastrous. Uncertainty about what was actually happening to the climate meant that people were unwilling to start taking drastic measures to reduce the consequences. But gradually the anxieties of the researchers were seized on by the media and communicated to a wider audience. In 1981 the greenhouse effect reached the front page of *The New York Times*. A couple of years later, a record-beating hot summer in the USA led to the greenhouse effect and manmade emissions being blamed. At the same time a spokesman for the National Academy of Sciences expressed his anxiety about the consequences of a warmer climate. He went so far as to say that we could have problems of a kind 'that we have barely imagined'.[4] The climate researcher Wallace Broecker used the metaphor that the earth's climate is like an angry beast that we have been poking with a sharp stick for some time.

Was global warming real, or were the sceptics right? The UN Intergovernmental Panel on Climate Change (IPCC), was established in 1988 to function as a scientific basis, a 'neutral intermediate link' between the climate sceptics and the climate radicals. Hundreds of experts summarized many years of work in 1990 and 1996. All available data indicated that there was a 'discernable human influence of global climate'.[5] One of the most important results of report no. 3, which came out in 2001, was documentation of the fact that the average temperature of the earth is increasing.

Just as important was the indication of the cause: 'There is new and stronger evidence that most of the warming observed over the last 50 years is attributable to human activities.'[6] There could no longer be any doubt that we humans are influencing the climate. The concentration of carbon dioxide in the air over Hawaii had by 2002 increased to 15 per cent above the 1960 level. The last series with the report from 2007 confirmed that it is 'very likely' that emissions of carbon gases can explain the measured temperature increase since the mid-twentieth century.[7] This conclusion is still doubted in certain scientific and political circles, despite the fact that new articles are continually being published that support the conclusions of IPCC.

Against the backdrop of increased emission of greenhouse gases, it ought already to be possible to notice the effects of global warming at certain places in the world. Can specific natural disasters of recent years be direct consequences of global warming?

CHANGES AND DISASTERS

Between 2000 and 2003 the British journalist Mark Lynas travelled to areas where global warming had already left its mark. In his book *High Tide* he tells of his travels and what stricken people in Alaska, China and the Pacific area felt were the causes of changes in the course of nature. Everyone he talked to had experienced minor or major changes in how nature was behaving – often right outside their front door. This could be extreme drought or a rise in the level of the sea or pollution of groundwater. Even global climate changes are first noticed at the local level, often by poor people with little chance of moving and starting somewhere else. *High Tide* is a report from a borderland where extreme weather reaches the point where it represents a climate in change. Over the past five years, similar reports have been more frequent in newspapers and periodicals. Extreme weather over the past few years has brought climate change into everyone's consciousness.

A new tendency is for the media to directly connect individual disasters to global warming, such as floods in Europe and hurricane disasters in the USA in recent years. But climate researchers, very interested in trends and average weather, point out that this is usually a dubious practice. Disasters caused by tropical hurricanes, such as Katrina, and Rita, which hit the USA in 2005, are examples cited. Hurricanes can, at worst, become stronger and more frequent as a result of global warming. While this continues to be hotly debated in research circles, most people agree that climate change leads to an increase in natural hazards. And this places some of the blame for natural disasters on the shoulders of Western consumers.

WHAT IS AT STAKE?

The effects of climate changes and more frequent natural disasters on today's societies cannot be overestimated. About 3.4 billion people, close to half the world's population, presently live in areas that are exposed to natural hazards. Compared with the 1990s, the number of people affected by all types of natural disasters has risen from 211 to 256 million per year. Since 1970 a total of 4.6 billion people have been affected worldwide, either in the form of damage to their houses and property, economic loss, evacuation, injury or, in the worst instance, death.[8] The increase is mainly due to disasters connected with too much or too little water: drought, floods and hurricanes. Droughts and floods are the natural hazards

In the face of such natural forces, human attempts at survival can appear to be useless. The 1991 volcanic eruption from Pinatubo, the Philippines.

that affect most people today and that will increase most as a result of global warming. Climate changes also mean that areas which previous only had few natural disasters are affected more frequently. The extent and consequences are dramatically larger in underdeveloped countries than in the rich north. Heat waves and severe cold will become a threat, especially for the elderly. Glaciers will melt and the level of the oceans will rise. Countries with a high population close to sea level will be particularly vulnerable both to a reduction of land areas, increased salinity of groundwater and a great risk of flooding when storms rage.[9] The number of environmental and climate refugees in the world is thought to be about 25 million[10] The World Health Organization (WHO) estimates that 77,000 people die annually in Asia and the Pacific region as an indirect result of climate changes.[11]

EUROPE WILL BE HIT HARDER

The summer of 2003 was exceptionally hot. Persistent heat was the natural hazard that led to southern Europe being hit by a disaster of unusual dimensions. Over 72,000 people died during a heat wave.[12] Just for once, Europe topped the disaster statistics. Because of the increased longevity of the population and a high proportion of elderly people, extreme temperatures now constitute Europe's greatest natural hazard. The countries round the Mediterranean are most exposed, and the 2006 disaster claimed the most lives in Italy, France and Spain. Since heat waves will become more frequent in Europe as a result of global warming, this is bad news. The number of those who will die as a result of natural disasters in Europe will therefore probably increase in the years ahead. Since 1970 2 per cent of all deaths resulting from natural disasters in this part of the world have been linked to heat waves. A total of 45 per cent of all deaths have been caused by natural hazards that may possibly increase in extent in the event of climate changes.[13] This also includes extreme cold. In the winter of 2006 a wave of cold weather took the lives of 800 people in the Ukraine. In the summer of the same year, extreme temperatures came once again to Europe, with 3,300 dying during the heat wave in July and August. France and Belgium were hardest hit. Europeans have been reminded by these disasters of two important things: first, that we are still vulnerable to natural hazards and, second, that climate change can cause an increase in both the frequency of natural hazards and in our vulnerability to them. From having been affected very seldom by geophysical disasters such as earthquakes and volcanic eruptions, Europeans face an uncertain future with climate-related disasters.

9

The Tsunami

'Earthquakes, as predicted in the hadiths, are among the
most important signs of the Last Day.'
Harun Yahya (pseudonym), Turkish author and theologian [1]

In the depths below Indonesia lies one of the big 'machines' that presses, squeezes and destroys the earth's crust. The Indo-Australian plate is being pressed northwards, beneath Sumatra and further downwards in the earth's mantle at a speed of 4–5 cm a year. This results in enormous friction and the building up of tension. Occasionally, when the rocks cannot take any more, a break and subsequent earthquake occur. So it is no coincidence that powerful earthquakes take place in the area. Furthermore, Indonesia is full of active volcanoes. There are over 70 of them. Southwards from Burma the islands and volcanoes are strung out in an arc as in a necklace: Andeman in the north, Sumatra, Java and finally Timor in the southeast.

From a larger perspective, the course of the Indo-Australian plate northwards is one of the most dramatic things taking place on the earth today. The collision with Sumatra occurs along the edge of the great Indo-Australian plate. The main seat of the drama lies further to the northwest, in the Himalayas. The climate, culture, ecology and topography of this part of the world have been formed by the – geologically speaking – brutal collision between India and Eurasia. On a map, the mountain chain almost gives the impression of having been formed by a snow plough that has got stuck in deep, wet snow. Everest lies there like the highest snowdrift piled up right in front of the plough – or India. The earthquakes in Kashmir (Pakistan) in October 2005 and Sichuan (China) in May 2008 were the most recent of whole series of quakes and disasters caused by the collision between India and Eurasia.

Sumatra has been described as a geologically highly danger-ous place. Millions of people living on the plate-boundary have literally felt this. One of the most infamous natural disasters of recent times was created by the eruption of the Krakatoa volcano in 1883. The volcano lies in a group of islands in the Sunda strait between Sumatra and Java. The explosion was so violent that it caused a huge tsunami that claimed the lives of 36,000 people. The author Simon Winchester has claimed that the disaster contributed to a Moslem apocalypse-motivated rebellion against the Dutch colonial powers. The eruption and the tsunami were interpreted as a sign that the Mahdi would come. Mahdi is a Messianic figure who will start a holy war against the infidels that will ultimately lead to the Day of Judgment.[2]

A whole series of extremely violent earthquakes and volcanic eruptions have put people to the test in this part of the world from time immemorial. As mentioned earlier, the supervolcano Toba lies north of Sumatra. The Toba eruption about 71,000 years ago is one of the most powerful we know of. Then there was the explosion of the volcano Tambora in the Sunda islands in 1815 that caused the deaths of more than 90,000 people. The consequences of the eruption could be noticed all over the world for several years. Both in 1833 and 1861 the group of islands was hit by earthquakes measuring over 8 on the Richter scale. But the biggest earthquake so far took place in the morning, local time, of 26 December 2004.

ANATOMY OF A DISASTER

It was one of the most powerful earthquakes ever measured, surpassed only by a few in Alaska and Chile. From the zone of fracture 30 kilometres deep off Sumatra, the fault opened up like a zip at a speed of up to 8,000 km/h. The fault raced northwards, slashing a 1,300 km-long zone of the earth's crust.[3] In Indonesia the quake caused considerable devastation. The released energy was colossal, equal to half the yearly consumption of the USA. On the Richter scale the earthquake measured 9.2. Eight minutes later, the quake was registered at the Pacific Tsunami Warning Center on Hawaii. From the epicentre the fault had shot up towards the sea bed, causing a vertical displacement of up to 20 metres. It was this modest movement that led to 26 December 2004 becoming a dark day in the memories of millions of people.

The pressure wave that arose from millions of cubic metres of water being pressed together moved at great speed in all directions. Thirty minutes later, the tsunami made contact with the coast of Sumatra, after two hours it reached Sri Lanka and after three and a half hours the Maldives. Wherever it approached land, the height of the wave increased. Those who were on the beaches or along the coast saw the waves tower up before they were consumed. Over 230,000 people in the area round the Indian Ocean died.

The earthquake could not have been foreseen. Nobody knows when major earthquakes will challenge our societies. But we know that they will come sooner or later. Earthquakes do not occur entirely by chance. They are, broadly speaking, confined to certain zones. The longer the time that has passed since the last earthquake, the more powerful the next one will be. A new theory proposes that the tensions released by an earthquake do not disappear but are stored in other parts of the faults. This in turn can lead to an increased danger of new earthquakes.[4] The theory is based on studies of aftershocks, which represent a third of all earthquakes measured. They follow a particular pattern, known as the Omoris Law, where the number of aftershocks falls to ten per cent after ten days and to one per cent after a hundred. By applying this principle to the main quake it has been discovered that the danger of new major quakes in other parts of a fault increase after a major earthquake. The theory has been successfully applied both in California and Turkey. After the earthquake of 26 December 2004, thousands of aftershocks were registered. The quake shifted tensions southwards and may have influenced the new earthquake that shook Sumatra on 28 March 2005. A further 300 kilometres of fault zone off Indonesia were activated.

The 26 December earthquake could not have been prevented. The tsunami was also impossible to predict. One could not know that the earthquake would cause displacement of parts of the sea bed. On the other hand, the journey of the tsunami towards land had been discovered by warning systems that might have saved thousands of lives. But the tremendous consequences of the tsunami could also have been reduced without the use of new technology. The solution is to reduce human vulnerability and the social causes of the extent of the disaster. The tsunami was such

'The Protector'. Drawing by Mufti Dinda, aged 13, from Aceh, Indonesia, who described the disaster: 'A huge earthquake put a sudden end to a lovely morning. The children from next door and I were playing hide and seek. I saw people running out of the houses in panic. Everyone gathered in the middle of the street – me, too. I heard many women call out "Laailaha illallah" [there is no god but Allah]. We never thought that the tsunami would hit us.'

an enormous disaster because it struck down vulnerable people. The poor had dwellings that were built to withstand neither tsunamis nor earthquakes. Sri Lanka and Indonesia have had other priorities than measures against natural disasters. Half of the families in the Aceh province live beneath the poverty line, despite the country being rich in natural resources. On Sri Lanka every third or fourth family lives below the line. Poverty, social unrest and civil war are among the societal roots of the disaster. Socially constructed vulnerability was the most important cause of the high death figures. A network for monitoring tsunamis in the Indian Ocean is necessary – and it has now been implemented, based on that which exists in the Pacific. But it is equally vital to improve conditions of life for millions of people in the coastal areas of South Asia.

SOMEONE TOLD US TO RUN

The experiences of those who followed the disaster at close quarters are dramatic. They tell us something of what we feel about suddenly having our lives turned upside-down. The greatest devastation and loss of human life occurred along the coast of Aceh province on Sumatra. 'Someone told us to run', 13-year-old Rahmianti relates.

> People ran to the hills, others to the mosque. Afterwards, my father went to see what had happened to our house. He came back and said, 'There is nothing.' My school was washed away with all my class mates. I don't know what happened to my friends. Only two of us survived. I want to go back to school so badly, but even if we sold all of our sweet potatoes we would not have enough money for the bus fare. Now we are living in a tent up here by our fields, on higher ground.[5]

Other people from Aceh talk about how the disaster has changed their lives for ever. Mahudin (age 60) had a career as Aceh's best-known comedian. 'Art is nonsense, it's nothing, trivial, worthless. I am going to be a preacher or something.'[6]

Mahudin was not alone in turning to God. For 29-year-old Sinta the disaster was an alarm call. She has now promised God to become a good Moslem and to pray five times a day.

> You have seen how the only buildings left by the wave in so many places was the mosque? That was God's way of telling us to be good Muslims. I truly believe that. It was a clear warning. Many people in Aceh were gambling. Students at the university were cohabitating and having sexual intercourse. We have a saying here that Aceh is the verandah of Mecca. I should have thought more about that earlier.[7]

God's role has been important for many people in Indonesia in their attempt to try and understand the disaster. The village leader Bunchia from Blang Me, where no children and few women survived, answers the question of why the village was wiped out: 'It was God putting us to the test. He is trying to make people better. If we are not tested, we can never improve.'[8]

2004 was the year when a natural disaster once again became global. It was the tsunami that brought the natural disaster to Europe. Despite the fact that the drama took place thousands of kilometres from our borders, the tsunami was possibly the first unifying natural disaster in Europe since the destruction of Lisbon in 1755. Europeans were brutally torn from their normally safe position as observers and transplanted into a reality where friends and families could be involved. In the days after the disaster it was feared that as many as tens of thousands of Western tourists on holiday in Asia had died. The human tragedies of the disaster poured in over us. Personal loss and the search for survivors and the dead dominated the news worldwide. No one could remain unaffected by all the sufferings. For many people in Europe as well as Asia, the disaster became personal. Major natural disasters can be difficult to grasp when one sees them from the outside and from a different culture. When some of 'us' are involved, suffering becomes personal.

In Asia complete villages were swept away. The WHO announced on 31 December 2004 that five million people were homeless in the affected areas. 1 January 2005 was declared a public day of mourning throughout the world. Before the year was over, it seemed as if this was one of the worst ever natural disasters. The news that streamed in from the disaster areas during the first week was tragic: survivors who desperately searched for their dear ones, looting, the sale of orphaned children, mental problems in adults and young people alike. The death figures rose from day to day. The need for emergency aid was acute. Large sums of money were collected by such organizations as Red Cross and Médecins sans Frontières. Western victims triggered the mobilization of the largest collection campaign the world has ever seen.

After the shock headlines of the media came analysis and reflection. Everyone tried, from their own standpoint, to understand the disaster, whether it was relatives, politicians, sociologists, geologists, psychologists, philosophers, clergymen or the man in the street. What had happened – and why? Would the disaster have long-lasting consequences? How should we tackle so much suffering? Could similar natural disasters take place in other countries – or at home?

In the weeks after the tsunami disaster everybody's eyes were on South Asia. The devastated communities fought to recover and to return to a normal life. In the meantime, the tsunami had changed how the world thought about natural disasters, poverty, technology and vulnerability. Was it poverty, the dangers of nature or a lack of technology that was the cause of the disaster? The tsunami was retold and interpreted via popular culture, the media, academics, musicians, actors and fellow human beings. It was as if the world had united in support of the victims of the disaster. This led to an enormous focus on natural disasters and poverty, both among politicians and people in general. The world's seven largest industrial nations promised after a meeting in London in February 2005 to reduce the debt of the world's poorest countries by up to 100 per cent.

At the same time, internal investigations were taking place in a number of Western countries to find out what had gone wrong with the handling of the crisis. Criticism by victims and survivors was mounting. Accusations of incompetence and a lack of flexibility and humanity rained down on foreign services in a number of countries after the disaster.

The country hardest hit was Indonesia, with 167,000 dead. This has left a profound trace on Aceh province, which lost the highest percentage of inhabitants. Many were afraid to return to their homes and were frightened of water, and had problems in tackling memories of the disaster, repeated nightmares, emotional numbness or a feeling of helplessness. It was estimated that half a million people would need psychological help and that 100,000 of these would be badly affected.[9]

After eight months' hard work in Aceh, some of the community's normal functions started to operate again. Most of the children were back at school though two thousand schools had been destroyed. But the process of reviving the northern parts of Sumatra and the island of Nias has cost a lot. Over 570 national and international organizations, the UN and many others have worked hard together with the survivors. More than 1,000 houses were built every month. But the challenges were still daunting a year after the disaster. About 67,500 people were still living in tent camps in autumn 2005, and even more had little chance of getting out of the dilapidated barracks that were erected after the disaster. The economy of the region was on a downturn and it was feared that a further 600,000 people might end up below the poverty line. The profits of the building boom were eaten up by inflation. The earthquake had also led to parts of Aceh sinking by one and a half metres. The coastline has changed and groundwater has been polluted.[10]

PERILOUS NATURE HITS BACK

No one could avoid hearing about the tsunami disaster. It became a major subject of conversation and the starting point for endless discussions and analyses. During the period from February to October 2005, the number of hits for 'tsunami' on Google rose by 35 million. The natural disaster set discussions in motion of which the world had not seen the like since 1755. Why did as many as 280,000 people die during the disaster? One of the most widespread explanations was 'perilous nature'. Nature had once again hit back at frail humanity. Everywhere in the world the focus was on the enormous power of the earthquake ('40,000 Hiroshima bombs') and the size of the tsunami. One of the most one-sided assertions that it is nature that creates natural disasters was printed in *Popular Science* in May 2005: 'Tsunamis, volcanoes, hurricanes, landslides – The single certain thing about nature's killers is that they will strike again, and again. Our only defense: ever better prediction and protection.'[11]

At the same time as the discussions about 'perilous nature' were resounding like some echo from the past, people's vulnerability was included. The greatest loss of life had occurred in areas with many poor people. Even though tourist destinations in Thailand were also hard hit, hotels were better equipped to withstand the earthquake and the tsunami than flimsy tin huts. The issue of poverty was high up on the list of priorities for reducing the extent of natural disasters.

The third type of reaction to the disaster had a completely different focus. It was claimed that the real cause of the disaster was human sin. Religious people from all corners of the globe used God to explain the tsunami.

Religious reactions to natural disasters today are often elusive. They are often limited to a sentence here ('God wanted to punish us'), a headline there ('Thought Judgment Day Had Come') and a few more or less curious accounts. Religious attitudes are seldom taken seriously by disaster researchers. This despite the fact that the religious dimension can be pivotal for an understanding of how disasters – the tsunami disaster, for example – affect us afterwards.

Reactions from Christians throughout the world came at once. In free-church and evangelical circles, it was claimed that the earthquake off Indonesia fits a trend of more and more frequent earthquakes. This will culminate in a great, devastating quake of biblical proportions. 'Why is there such a dramatic increase in the age in which we now live?' one minister of the Pentecostal Movement rhetorically asked. 'It is very stupid indeed not to take these signs [natural disasters] seriously – extremely seriously. There are far too many signs for these to be rejected as coincidences . . . We who are alive today are living in prophetic, highly dramatic times.'[12]

This interpretation of natural disasters is linked to the biblical Fall, the lesson being that it is difficult to escape. 'That means that such disasters are not directed against those who have been affected, as a form of personal punishment, but belong to the curse that struck the earth because of man's rebellion against God in Paradise.'[13] A few people have focused on the enormous consequences the Flood in the Bible must have had when compared with the tsunami disaster, which, despite everything, was much smaller: 'Just imagine what the tsunamis of the Great Flood did to our planet when the waters were higher than the highest mountain.' This type of 'apocalyptic disaster' is interpreted as teaching us that we must take care of each other and that no one is to live on this earth for ever: 'We live on an earth that is corrupted and accursed because of what is referred to in the Bible as the Fall.'[14]

A link between the disaster and God was also maintained by some Catholics. A priest from Sri Lanka claimed that the disaster was 'a message from Christ to turn away from a sinful life. Norway and other Western countries can be the next victims if a change does not take place in people's relation to God.'[15]

Vrbs Rhodia labefacta: Terremotu.

How could God permit such destruction? Clergymen gather together for prayer in an open square in Rhodes after an earthquake in 1481.

In the USA, conservative Christian organizations expressed concern but also hope regarding the possible significance of the disaster. The leader of the Institute of Creation Research outside San Diego in California, John Morris, pointed out – as did many others – that the tsunami of South Asia was small when compared to that of Noah's time. The time of the major disasters may soon be over: 'We are waiting full of hope for the time when such disasters will cease, and the earth once more become new . . . Come soon, Lord Jesus.' Others were more sceptical. How could one distinguish between a 'normal' natural disaster and one that also conveys a message from God? One of the answers received was that God will always spare the innocent. Since the tsunami disaster struck such a wide cross-section of humanity, it was probable that many pious and good people were among the victims. So the disaster could not have been caused by God.[16]

KARMA

Many of the 16,400 people that perished along the coast of India were Hindus. In Hinduism the question is not asked as to why the gods allow disasters. The gods' ways are unfathomable, but they are always good. An important point about Hinduism is that the life one leads will be able to be affected by the good or bad things one did in a previous life: karma. Hindus do not believe in salvation or a life after death but in reincarnation and an eternal, unchanging soul. One is reborn until one reaches the final goal: release from the realm of reincarnation. For then one is united with the Absolute, and one's karma erased.[17] The Hindus who died during the tsunami might have got a bad reincarnation, some people claim, because dangerous spirits are associated with the sea. Furthermore, a violent death can affect the soul in a negative way.

The tsunami in South Asia also meant that the question of 'God's role' in our sufferings reached Buddhists. Karma is also important to Buddhists when trying to explain the causes of suffering in the world. Buddhists in Thailand or on Sri Lanka probably did not believe that the enormous loss of human life during the tsunami was due to the actions of those who died. For Buddhists it was more important to focus on helping the survivors. By helping other victims of the disaster, Buddhists can improve their own karma. Furthermore, Buddhists believe that one finds oneself in an intermediate state when one dies. Here, those who can remember something of their own death are able to influence their karma by turning their own suffering into a sacrifice and thus into something positive. When one shows concern for others, this leads to a good reincarnation.[18] Some Buddhists believe that one can protect oneself against accidents and natural disasters with the aid of rituals with texts that have magical powers (*parittas*).[19] They can be read out both to prevent disasters and to commemorate the dead. After the tsunami of 2004, *parittas* were broadcast over the radio and read aloud in commemorative ceremonies. *Parittas* were thought to be capable of calming the spirits of the dead. Both in Thailand and Sri Lanka there were innumerable reports of ghosts and spirits in the time after the disaster. The local population kept away from the beaches out of fear.[20]

It was not the adherents of Christianity, Hinduism or Buddhism who were hardest hit by the tsunami. The vast majority were Moslems. No country has more Moslems than Indonesia – no less than 87 per cent of the total of 225 million people. It was also Indonesia that suffered the largest loss of human life. Are the explanations provided by Sinta and Bunchia from Aceh that Allah was behind the disaster mere superstition, or does this interpretation of natural disasters have a basis in Islam?

IS ALLAH GOOD?

Najeeb-ur-Rehman Naz is dressed in a traditional white robe and sitting at his computer in the spartanly furnished office of Norway's most imposing mosque. He is an imam at the World Islamic Mission in Oslo. The community was established in the 1980s and today it has about 5,000 members. As one of the 'guardians' of the faith, especially for Pakistani Moslems, he knows well how natural disasters are interpreted by Moslems. A starting point for an understanding of Moslems' attitude to disasters is to be found in Sura no. 99 in the Qu'ran. Its title is 'The Earthquake':

99:1 When the earth is shaken to her (utmost) convulsion,

99:2 And the earth throws up her burdens (from within),

99:3 And man cries (distressed): 'What is the matter with her?'

99:4 On that Day will she declare her tidings:

99:5 For that thy Lord will have given her inspiration.

99:6 On that Day will men proceed in companies sorted out, to be shown the deeds that they (had done).

99:7 Then shall anyone who has done an atom's weight of good, see it!

99:8 And anyone who has done an atom's weight of evil, shall see it.[21]

'The sura talks about events preceding the Day of Judgment', Naz explains. 'Then, major earthquakes will take place. The whole universe will collapse. Everything will be laid waste. Everything will pass away.'

Seriously and with deep concentration, Najeeb-ur-Rehman Naz explains the divided view of natural disasters in Islam.

> Most people interpret natural disasters as either being a test from Allah or a sign that the Day of Judgment is near. God tests us by imposing suffering on us, to see what we do in extreme situations. Do we still remember God? Thank God? Worship Him despite what has happened?

Moslems recognize physical nature as something independent and outside belief that can be understood with the aid of science. At the same time, the world is governed by Allah as the prime mover. So it is not only in conservative Christian circles than the view of natural disasters is divided.

> When it comes to earthquakes, they come from the depths of the earth. But it is God who sets them in motion, even though they can be physically explained. An earthquake does not necessarily lead to a disaster. The final, decisive power is with God alone. According to our faith, God can do everything. In 1999 alone, over 20,000 earthquakes were registered. In 1999 alone!

The mixing of religion and science has for a long time been a delicate theme in a number of Western countries, especially in the USA. There, creationists have involved themselves in the teaching of geology and biology in schools, and have pushed for the teaching of 'intelligent design' rather than evolution. Researchers and parents have protested violently. It has seldom been commented on that conservative Christians and Moslems have much in common when it comes to how they view science – especially the theory of evolution. In Pakistan, for example, Allah's role as creator is part of the biology curriculum, while in the USA it takes court cases to keep the biblical story of creation out of the teaching of natural science. Among orthodox Moslems the thirst for knowledge is primarily motivated by the need to understand God better.

Religious explanations of natural disasters are not a phenomenon that exclusively belongs to learned theological circles. Moslems, in both cities and countryside, share the same attitudes. After the earthquake in Kashmir in 2005 Allah's punishment and testing were repeatedly advanced as an explanation. Similar attitudes have been documented in Moslem countries after other natural disasters. After an earthquake in Egypt in 1992, many people claimed that it either wholly or partially had a religious origin linked to the Day of Judgment. A 40-year-old man told the social anthropologist Jacqueline Homan: 'The earthquake was related to God.' And why had it occurred? 'It was sent as a memorandum from God to show how strong He is in both place and time.' The co-existence of a religious and scientific world picture is clear in this statement by a male student in Cairo: 'Everything comes from God, only He can save our souls, *but* you can also explain earthquakes scientifically . . . There are scientific controls on the environment and earthquakes and so on are natural events, but they are ultimately controlled by God.'[22] There is apparently no opposition between faith and science. The scientific and the religious world picture exist side by side.

In Shia Iran, we have seen examples of drought also being interpreted as God's work. The author Torbjørn Færøvik encountered exhortations to pray for rain at a border station to Iran in 2000:

Iran has been hit by the worst drought in human memory. People are suffering and the animals are dying. Whole flocks of sheep have starved to death. Goats, donkeys and mules. Even the camel is shaking at the knees. The rivers have dried up, and the irrigation systems have become ravines of dust. Now is the time for praying. All inhabitants of Iran must pray to Allah![23]

THERE CAN BE NO DOUBT

Is it possible to understand the sura about the earthquake and other accounts of disasters in a symbolic way? Najeeb-ur-Rehman Naz sweeps aside all doubt.

There are no gradations of faith. The Koran is the word of God, the word of the true God, and what it says there cannot be doubted. This is something one has to believe. Furthermore, all interpreters of the Koran agree that the sura about the earthquake applies to events leading up to the Day of Judgment. There may admittedly be small differences among those versed in the Koran, but this applies first and foremost to areas that have to do with rituals and traditions – not with the content and fundamental principles of the Koran.

Natural disasters are mentioned in a number of the recordings of the tradition of the Prophet Mohammed, or *hadith*. The prophet foresaw that earthquakes would become more frequent before the Day of Judgment. The disasters will culminate in The Great Earthquake, which will swallow up the entire world in an inferno of dust, flames and smoke. For Moslems there is nevertheless hope. Everyone will be given new life and subsequently called to account. But before then, one has to pass through the Day of Judgment. And Naz repeatedly stresses that the Day of Judgment *will* come.

Belief in the Day of Judgment is a prerequisite for being a Moslem. There is great emphasis on this in the Koran. One has to think a great deal about life after death because it is to be eternal. It is an absolute demand that one believes in the Day of Judgment.

The extremely conservative Turkish Moslem author 'Harun Yahya' is among those who have written about the view of natural disasters in Islam.[24] He has found a number of contemporary signs that the Day of Judgment may be closer than we believe. Just in the 1980s a whole series of events have been interpreted as precursors of Judgment Day. All of them have apparently been predicted by the prophet Mohammed: the change of regime in Iran, the war between the two Moslem nations Iran and Iraq, 20,000 people killed as a result of the gas emissions in India (Bhopal), the earthquake in Mexico City (1985) where 10,000 people died, the eruption of the Nevado del Ruiz volcano, the floods in Bangladesh that claimed the lives of 25,000 people, floods in Rome, forest fires in China, assassination attempts on heads of state, the hurricane in Edmonton in Canada (1987), the spaceship Challenger that exploded in 1986, the AIDS epidemic that put an end to the sexual revolution, the Chernobyl disaster, the hole in the ozone layer, special planet constellations in 1982, the fall of the Berlin wall, the demonstrations in Tiananmen Square in China, the conflicts in Bosnia and Kosovo, the Ebola virus, Halley's Comet, countless earthquakes throughout the world, hurricanes, tornadoes and storms that resulted in thousands of deaths, increasing poverty, starvation, the collapse of moral values, prostitution, infidelity and homosexuality.[25]

Harun Yahya stresses that many people will claim that natural disasters do not have anything to do with God. But that is only because they lack awareness, he asserts. The interpretation of natural disasters seems clear. 'By the will of Allah, an earthquake may happen at any time.'[26] Allah can intentionally create uncertainty and instability in parts of a country. God actually creates disasters for special purposes. They are warnings. 'Allah creates these disasters to show us how insecure our habitation can sometimes be. These outbursts of nature are reminders to all mankind that we have no control whatsoever over the planet.'[27] It is in that way Allah shows that this world is only a temporary place. A place where we are constantly being tested.[28]

'We are living in the latter days.' Everything in the entire universe will be destroyed.

With this background material, it is easier to understand the reactions of Moslems hit by the tsunami in 2004. A leader of a radical Islamic group in Indonesia believed that the disaster was due to the fact that people had abandoned Allah. 'They were not true to their faith. Allah had given the Acehnese Islamic law and they did not implement it . . . It is crucial that the survivors, and indeed all Muslims, understand that this was a warning from Allah. If they don't become true Muslims then they will be struck down.'[29] A woman from Banda Aceh concludes in this way: 'We have been punished by God because we did not pray enough. Look at the mosques – they are still standing.'[30] This is not superstition, as many in the West would claim. It is faith.

Contemporary history comes into being while we are about our everyday business. It is often difficult to find milestones or trends in developments, even after a dramatic event, because we are too close to them in time. Furthermore, the effects of a major natural disaster can last for years or decades and many of the recollections can be of a personal nature.

'This is the moment to have serious conversations and to find common ground for a peaceful solution in Aceh.' The statement came from Malik Mahmud, the exiled head of the province of Aceh, in February 2005.[31] The basis for this was that the tsunami disaster had brought about a softening of relations between the Indonesian authorities and the Aceh Liberation Movement (GAM). The conflict has been going on since 1976, and it has been bloody. The breakthrough came after negotiations in Helsinki in August 2005. A peace agreement in four stages was to ensure peace in the province. Combined with a softer line from Indonesia's president Yudhoyono, the disaster helped change the focus from war to dialogue. 'We would not have come here without the tsunami . . . It showed us that there has been enough suffering in Aceh', as a leader in the province put it.[32] In the course of 2005 Indonesian security troops withdrew from the province and GAM handed over its weapons.

10

Perilous Nature

'And every living substance was destroyed which was upon
the face of the ground, both man, and cattle, and the creeping
things, and the fowl of the heaven; and they were destroyed
from the earth.'

Genesis

A milestone within international disaster research has been the real-
ization that it is not necessarily 'perilous nature' that causes natural
disasters. They are the consequence of the *interaction* – or lack of the
same – between nature and society. Natural disasters are just as
often the result of how our societies are organized as of geophysical
forces or of the extreme weather conditions that trigger them.

This relatively recent insight has overturned the traditional
view that humans are and remain the powerless victims of natural
forces. Particularly within academic circles, attitudes have changed
dramatically since the 1970s, although outside specific academic
circles 'mighty nature' is still the usual scapegoat when disasters
strike – as we saw after the tsunami disaster of 2004.

ARE NATURAL DISASTERS NATURAL?

The natural disasters of the 1970s had widespread and unsuspected
consequences. Drought and famine in Sahel, major earthquakes in
Nicaragua and Guatemala, cyclones in Bangladesh: a number of
researchers sought answers as to why natural disasters particularly
hit poor countries so hard. Had the hurricanes become more powerful or the earthquakes more frequent? Apparently not, was the
answer. Despite this, new overviews showed that the extent of
natural disasters had escalated since the 1950s. Were we too badly
protected against 'perilous nature'? During the 1970s several people
felt that it was high time to reassess much of what we thought to
be known about natural disasters. A group of researchers with a
background in developmental studies pointed out that vulnerability to natural hazards was closely linked to underdevelopment

and the marginalization of people in poor countries. The conclusions they came to were that natural disasters are not primarily created by nature but arise as a result of an unequal distribution of the earth's resources.[1]

One of those who was inspired by the new ideas concerning natural disasters and vulnerability was Kenneth Hewitt. He was in the process of establishing himself as a researcher at the university in Waterloo in Canada. In 1977 Hewitt organized a seminar on disasters in order to discuss the new insights. The seminar in Waterloo brought together an untypical group of disaster researchers. Most of them were geographers who had focused on natural disasters in underdeveloped countries, with drought or famine as their speciality. One of the aims of the meeting was to combine radical political and sociological ideas with ecological and the scientific ones.[2] The meeting was an eye-opener for Hewitt and the other participants. A short while after the seminar in Waterloo, Hewitt met the geographer Michael Watts. When the latter heard that Hewitt was planning to publish a book of the seminar in Waterloo, he wanted to make a contribution. Hewitt was challenged to write a new preface, in order to sound out new disaster research. For his part, Watts was in the process of completing a pioneer piece of work on famine in Nigeria.[3] In 1983 the book of the seminar, along with the new ideas, was spread to the academic world under the title *Interpretations of Calamity*.[4] Hewitt's preface became a classic for a whole generation of researchers.

The book caught people's attention and gradually reached researchers outside the field of geography. Much of what people thought they knew about the causes of natural disasters had to be revised. With *Interpretations of Calamity* ideas that had crystallized in the 1970s matured and grew. Hewitt spearheaded the new change of attitude. What was so new and radical about it? Hewitt claimed that we humans are the most important cause of natural disasters taking place, at least in countries with widespread poverty. We are the ones who must reduce the vulnerability of society to hazards. The meeting in Waterloo helped shift the focus away from natural forces to the vulnerability of humans and societies. Their conclusion was that natural disasters in underdeveloped countries can best be understood via a social science approach, not an approach based entirely on natural hazard studies.

The major interest in natural disasters among today's researchers is a relatively new trend. Hewitt has pointed out that even if our societies have always been hit by natural disasters, today's scientific understanding of them is young. The early beginnings can be traced to the American researcher Gilbert White in the 1940s. His aim was to reduce the damage caused by flooding in the USA. A number of other researchers became interested in disasters during the 1950s and '60s. The Second World War also led to Americans wanting to know more about what traumas could be caused by wars and technological disasters. So what could be more natural than to study the victims of natural disasters? At the same time, many people were very interested in scientific and technological solutions to protect us from natural hazards. This generation of disaster researchers gradually came to be known as the Chicago School and was to exert considerable influence on how disasters were dealt with in a number of countries, by the UN and in the academic world. The main message was that one can *adapt* to natural forces in such a way that the extent of the disasters is reduced. The means of doing so were a mapping of the risks, a monitoring of nature, the drawing up of evacuation plans and the development of new technology. Knowledge of nature and the development of the earth was seen as being the most important factor, because natural disasters were considered inevitable consequences of the extreme processes of nature. By transferring Western knowledge to underdeveloped countries, it was argued that many of the problems would be solved. This approach had proved a great success in rich countries – even if disasters in the West are often extremely costly, the number of deaths had been considerably reduced. However, this is not the first approach to take to reduce the scale of natural disasters in poor coutries.

The technology strategy was central during the UN's international decade for the reduction of natural disasters, launched in 1980. An ideological split opened up after the ideas of the Waterloo meeting had been spread.[5] Hewitt and his colleagues were labelled 'the radical critics'. They had passed a crushing judgement on the 'dominant approach' of the Chicago School.[6]

The main point of 'the radicals' was that if one is economically or ecologically marginalized, a transfer of technology will not help when disaster strikes. Measures must come from the grass-roots level of the exposed society. The vulnerability of people and institutions can be more decisive for the extent of disasters than the physical forces of, for example, a volcanic eruption. In other words, the technological angle of the Chicago School was not the best way of tackling the problems. They must be seen as part of a whole – the north/south problematics. Local knowledge, rather than imported Western experts, ought to be used to reduce the consequences of natual hazards.

Did the theories of 'the radicals' gradually become the dominating ones? A large number of researchers and others who work on problems connected to drought, famine and floods do so within the framework developed after the Waterloo meeting. Today, radical criticism is also used to understand the extent of natural disasters resulting from volcanic eruptions and earthquakes. Physical factors, geology, ecology, poverty, politics, technology, ideology, culture, religion, education, economy and institutional factors all help determine why hazards develop into disasters. Natural disasters have become social phenomena with roots in the processes of nature, especially in vulnerable societies. Even so, 'perilous nature' is still an extremely common explanation for why natural disasters occur – among researchers, politicians and in the media ideas originating in the Chicago School are still widespread.

Radical criticism has developed since the 1980s. Kenneth Hewitt continued work on his theories during the 1990s and he claims that both the technocratic and the vulnerability-oriented views of disasters define the non-Western part of the world as 'weak, passive and pathetic'.[7] During recent years, especially after the terrorist attacks of 11 September 2001, a change of attitude – according to Hewitt – has taken place in the way natural disasters are viewed in North America. People are more concerned about homeland security than before and spend great amounts of money combating an enemy now known as *risk*. The objective is a society without danger and there is a new wave of technological solutions on the way in. Vulnerability is changed into something people believe can be dealt with in the same way as poverty or 'perilouss' nature – something that can be combated with the aid of Western technology.[8]

The historian Greg Bankoff is one of those who has criticized the new concept of vulnerability longest. He claims to have uncovered an ideological basis underlying the concept 'vulnerability' that makes disaster research seem as exciting as the best detective story. The salient position 'vulnerability' has acquired within the field of natural disasters is a new Western invention, Bankoff points out. The worrying thing is that 'vulnerability' can turn out to be an ideology that makes it easier to maintain our control over non-Western societies and their resources. From this point of view, 'vulnerability and natural disasters' have taken over from post-war 'relief aid' and 'reduction of poverty' as the most important ideology for justifying Western intervention in many parts of the world, Bankoff claims.[9]

MODERN MYTHS

In spite of the many differing views of natural disasters, it is time to explode some of the commonest myths. One of the most widespread is that natural disasters are unique events that lie completely outside what is normal. One of Kenneth Hewitt's trump cards was that natural disasters, on the contrary, represent the *normal*. They are natural consequences of how we as humans act towards nature and the environment. Disasters and the resulting damage are *characteristic* of the place where they occur, not unexpected. This view is supported by disaster statistics. On average, 450 natural disasters take place each and every year – and that is not including the smallest ones.[10] But the usual attitude to natural disasters is that we can relax after one has struck, because it is improbable that it will occur again. But the next time a disaster devastates a society, just as much chaos and confusion will ensue, and we will be just as surprised as the last time. Few mechanisms enable the passing on of knowledge about natural disasters at the local level. Even dearly bought experience can quickly be lost; people disappear from the system without their knowledge and experiences being passed on.

It is more important than ever to try and understand earlier disasters and their societal consequences. Only the future can provide the key to how the past is to be interpreted. The way in which we understand the present and the future will continually affect our rewriting of knowledge about the past. One example of this is that research into disasters in the eighteenth and nineteenth centuries has taken account of people's vulnerability to natural forces. This is a result of new knowledge, not about the past but about the present. In addition, focus has shifted from the major events and the important people to mentalities, attitude and the apparently minor events. Popular attitudes and 'irrational' views of how the world and nature work often emerge after natural disasters. But historians have only used this to a small degree as 'windows' for understanding the present.[11]

There is a widespread belief that many types of natural disasters 'take no account of class divisions'. But time after time, after earthquakes or floods, it is people on the lowest rungs of the social ladder who have to bear the brunt. Natural disasters never strike blindly. The poor and weak in resources experience the greatest loss of life and material goods – they are worst insured, have the most mental problems to deal with and find it hardest to fight their way back to the normal rhythm of life. Rich and democratic countries, for example, have never allowed drought to develop into famine. India has not been hit by major famine disasters since the country became independent in 1947.

NATURAL DISASTERS AND HOW THEY INFLUENCE US

Rachel Carson's book *Silent Spring* (1962) changed our view of the state of nature. It was an eye-opener for many people to discover that our activities were in the process of destroying the earth, its cleanness, ecology and 'health'. Furthermore, Carson pointed out that the pollution of the planet ultimately rebounds on us. A similar important change in our view of the relation between nature and humanity occurred in the 1980s, triggered by the dawning awareness of global warming. We are in the process of changing the course of nature. For that reason, more of the natural and environmental disasters of tomorrow will be manmade. Combined with the criticism of 'the radicals', it is becoming clear that our dominance over nature is enormous – and is rebounding on the human race. The ecologically aware writer Bill McKibben has pointed out that we 'as a race turn out to be stronger than we suspected – much stronger. In a sense we turn out to be God's equal – or, at least, his rival – able to destroy creation.'[12] Humans move more earth and stones on the surface of the earth than nature itself manages to do with all its rivers and glaciers.[13] We have also changed the composition of the atmosphere. The concentration of CO_2 has increased dramatically since measurements started on Hawaii in the late 1950s. The climate is changing. Gone is the feeling of living on a planet beyond human control and a nature that remains unchanged and unaffected by our interventions.

When the division between natural and man-made disasters becomes even more erased, the former view of nature as dark, threatening and potentially deadly will became even more frightening – even though it only shows *its* vulnerability to our interventions. It is not volcanoes that we fear most nowadays but the total effect of natural forces and our own interventions. Now it is *nature* that is vulnerable to violent human forces. 'We can no longer imagine that we are part of something larger than ourselves', McKibben claims.[14] Now it is up to us to decide what is to happen. The future of nature and the earth now lies in *our* hands. Since we can exert a decisive power over nature and have the ability to control and reverse the course of events, we are able to slow down and stop the changes to the earth's climate. Man stands at last fully conscious at the summit of 'creation', until natural disasters knock us off our perch once again and remind us of our own vulnerability.

The attitudes we have to the nature that surrounds us are extremely complex today. Social scientists even find it hard to agree on what 'nature' is and what relation it has to our culture. But at the mental level things are different: We still have a strong link to ideas and mentalities many people consider to be over and done with. Religious explanations of natural phenomena and natural disasters are often claimed to be a thing of the past, for knowledge of nature has – despite everything – been the province of natural science since the mid-nineteenth century. The phenomenon is marginal, many people claim, almost a curiosity: 'We are after all modern, rational human beings!' But in most cultures and in many layers of society mythical and religious explanations of the processes of nature flourish – and of why people are hit by natural disasters. Even so, one can be seen in the West as ignorant or superstitious if one reacts by bringing God in after a disaster.

In his 'radical' criticism, Kenneth Hewitt claimed that this is a characteristic trait of the Chicago School's attitude to disasters – God has no place in a technocratic world view. But this school did not only remove God's role in natural disasters. It also removed human intervention. As Hewitt pointed out, traditional religious explanations had a particular view of the relationship between God and man: the problems were usually based on our own actions.

That natural disasters are often given religious explanations is something most researchers are well aware of. People often explain their losses after disasters within a religious framework, even though they come from cultures where scientific theories are well known. Hewitt points out that the further removed victims of disasters are from the urban, industrialized world, the more certain it is that disaster studies describe these people as fatalists, subjective and the victims of mystical, irrational and unscientific conceptions. This is now in the process of changing, even though religious historians on the whole are conspicuous in their absence. The British geographer and disaster researcher David Chester is among those with an interdisciplinary approach – he claims that a recognition of local religious explanations of natural disasters is absolutely vital if one wants to reduce vulnerability to disasters.[15] To improve prevention, crisis management, evacuation plans and recovery, local culture and religion must be taken into consideration. When people are pushed to their very limits, it is often religion that is left as their only hope. Religious leaders have an important function in giving believers support and protection during and after disasters. Religion had an important therapeutic role after the tsunami disaster of 2004.

If one believes that disasters are due to God's will, how does this affect one's attitude towards prevention? Surely it is useless to try and fight against something that comes from God Himself? One cannot oppose the will of Allah. 'What are we to do? Everything that happens to us is the will of Allah and we must submit to it', people say in Bangladesh. Most of those affected by annual flooding and cyclones in Bangladesh are Moslems. On the other hand, people have always done what they can to protect themselves against natural hazards, and it is only when their own measures are insufficient that they resign themselves. Furthermore, gender plays a role in whether one wishes to survive a natural disaster or not. In Bangladesh Pakistan, Indonesia and a number of other countries there is a preponderance of women and children that lose their lives. In Bangladesh, the problem was that women refused to use the new shelters that had been built as protection against cyclones. Cultural factors, such as Moslem *purdah*, meant that women felt uneasy about seeking shelter in places where there were unknown men.[16] So the 'battle' against natural forces is often fought in the cultural arena – with religion as a backdrop.

Can natural disasters lead to cultural and social changes? The social anthropologist Susanna M. Hoffman has the following cryptic answer to the question: 'no, but also decidedly yes.'[17] The answer is 'no' because people and cultures are often capable of resisting pressure from outside. They cling hard to traditions and ingrained habits. Disasters can therefore serve to strengthen already existing attitudes in a society. The answer is 'yes' because cultures are constantly undergoing change and are open to new impulses and ideas. In a way, natural disasters are like natural experiments to see how people and societies react to sudden, dramatic changes. In the chaos that arises during a disaster, everything is changed. One's normal rhythm is broken, and the most important thing becomes survival and the protection of the family. A number of aspects of our societies are revealed once the dust after the disaster has settled and the wheels of society are set in motion once more. The way in which society is organized – class divisions, priorities of authorities, alliances and ideologies – becomes visible. Action can lead to changes. Few events can help mobilize people more strongly than disasters. New leaders emerge. Existing ways of dealing with problems are criticized. Countless questions of both a practical and metaphysical nature come to the fore and call for answers. Myths and narratives are created, and religious questions and God's role discussed. Both religious and scientific explanations of causes are introduced. Processes that had already started before the disaster can be accelerated and create changes. Major disasters can signal that something is completely wrong with the priorities authorities have (infrastructure, poverty, understanding of risk) and can speed up criticism and the demand for reform. This can be important during the first phase of the disaster, when those affected are dependent on relief aid, as after the earthquake in Pakistan in 2005, where almost three million people were in danger of starving as a result of difficulties in procuring relief aid. The idea of a strong and all-powerful state can receive a serious blow after a major disaster. Investors can pull out if there is a risk of further disasters. A major disaster can set a country back many years when it comes to development. Ultimately, events can lead to changes of attitude and new ways of viewing oneself and society. This can give rise to permanent cultural changes.[18]

A number of factors influence how radical post-disaster changes can be. Were they the result of simple adjustments to an already antiquated pattern, or were entirely new processes instigated? Who was influenced? Were the changes comprehensive or limited to a small group? How large were the changes and how long did they last? A summarizing answer is that the degree of influence depends on the extent of the damage, how many people were influenced – and on how complex the culture is. Cultures undergo constant change, but the changes after a disaster can take place over a long period of time and can often be small and subtle.

Practically all victims of disasters need to create their own rituals and relics. These may be images of how things were prior to the disaster, objects found in the ruins, or places that become sacred or function as temples. If the loss has been too great, victims of disasters may have problems in relating to anything other than the past. This can result in their living in an isolated world, like the one prior to the disaster. It can take several decades for bad memories of natural disasters to lose their hold, and in many instances people live with the disaster for the rest of their lives. Victims may also be left with a strong feeling of solidarity, of having something in common with others that nobody else can share. Life often gains a new meaning – and they are able to feel more free and less bound by the material world.[19]

The attitude that much of natural disasters has human causes will become more widespread in the years that lie ahead. This may also lead to some of the psychological reactions changing. Stress reactions such as irritation, fear, insomnia and apathy are extremely widespread among those who have experienced natural disasters. It normally takes a fairly long time to deal with bad memories after a man-made disaster, precisely because there are scapegoats and people who can be accused – the disaster could have been avoided.

THE END OF THE WORLD

Just how dramatic social and cultural consequences can disasters have? Can they lead to entire cultures disappearing off the map? Can what happened on a small scale after the Parícutin eruption in Mexico do likewise to a whole people? If we look for traces of ancient civilizations that have disappeared, much would seem to indicate that climate changes and environmental crises have played a key role. The Mayas in Latin America and the Minoan and Mycenaean societies in the Mediterranean region are examples of cultures that disappeared. One of the boldest attempts to link natural disasters with changes to society has been made by the British journalist David Keys. In his book *Catastrophe* he claims that a volcanic eruption in the year AD 530 had enormous global consequences.[20] The actual volcano has not been localized, but Keys believes it must have lain somewhere in Indonesia. The eruption may have led to a global cooling that lasted for several years. Harsh winters and bad summers may then have led to flooding and epidemics, something Keys has found mentioned in historical sources. But this is only the beginning. Keys also presents a whole series of possible cause and effect relationships that include the rise of Islam and the fall of Chinese dynasties. On the basis of his findings, Keys predicts that a gigantic volcanic eruption in our time could lead to the death of hundreds of millions of people and to a change in the geopolitical balance of power in favour of the Third World.

Natural disasters are not need for cultures to be laid in ruins. The researcher and writer Jared Diamond has pointed out that a number of different conditions can affect how a society gets through crises, and that the most important one is how it solves its environmental problems.[21] Diamond uses examples from a number of cultures that disappeared. What, for example, was the person who chopped down the last tree on the Easter Islands thinking of? A common trait of societies we know have collapsed is that they were creative and advanced – not primitive, as might easily be supposed. Diamond's most important factors for collapse include damage to the environment, climate changes and the relationship a country has with its neighbour. If one refuses to take the environment, ecology and the interaction with nature seriously, it can have major consequences. Things are different for us today. We are a part of a globalized world. Major changes in one country can result in unimagined consequences for the rest of the world. But via our negative impact on nature (pollution, deforestation, non-sustainable development, climate changes) our own vulnerability will increase.

Prophecies concerning the last days of the earth do not come exclusively from strongly religious people. They also come from researchers, often in conjunction with the search by the media for the extraordinary and the frightening. Natural disasters rouse our curiosity, even though they also can mean our death. Proficient and inventive researchers have today tried to find most disaster scenarios. Natural hazards are interpreted on the basis of the earth's history and projected onto our own age. Meteorites and threats from outer space, supervolcanoes, enormous earthquakes, tsunamis, gigantic landslides. It is as if everything that has ever happened in the way of disasters can strike us at any time – at any rate in the course of a generation.

Europe's only active mainland volcano, Vesuvius, obliterated the cities of Pompeii and Herculaneum when it erupted in AD 79. The eruption was portrayed 'live' by the historian Pliny the Younger and, together with the casts of humans caught by the eruption, the natural disaster has gone into history as one of those that will never be forgotten. In the seventeenth century, too, glowing pyroclastic flows took the lives of several thousand people on the sides of the volcano. Can the volcano create a 'new' Pompeii for the archaeologists of the future?

Vesuvius still constitutes a risk. The last eruption in 1944 ended with a total of 26 deaths and the village of San Sebastino was laid in ruins. Its location on the attractive Bay of Naples, with good soil and many tourists, increases the risk of a disaster in the event of a new eruption. Today, houses lie on top of the lava that came from the volcano in 1944. Three million people could be affected if Vesuvius once more spews out ash and its volcanic

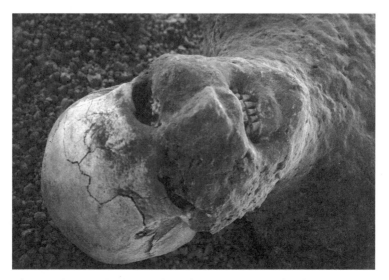

The Vesuvius eruption in AD 79 buried two cities. Pompeii is the most dramatic monument we have of a disaster. It also furnished us with much knowledge about life in Italy 2,000 years ago.

bombs. Hundreds of thousands of people live in the most exposed zone, close to the volcano. Despite the authorities' early warning plans, disaster geologists have published a depressing assessment of the safety of Naples's inhabitants: 'The rapid and successful evacuation of 600,000 people would be an extraordinary logistical feat at the best of times. In the midst of a crisis, against a backdrop of a rumbling volcano and in the face of stoical resistance by the population, it would seem to be doomed to failure.'[22]

In popular culture, the apocalypse has already taken place countless times in films and serials. In the real world, it can come tomorrow or in a thousand years' time. Nobody knows. At the same time, the events that take the lives of most people today are not linked to any of this but to hunger, poverty, drought and flooding. But we are at the hub of history, the point around which everything turns. Anything can happen. 'In fact, the Earth is an extraordinarily fragile place that is fraught with danger: a tiny rock hurtling through space', as the disaster geologist Bill McGuire has pointed out. The superdisasters can strike us 'at any time' and sweep us off the surface of the earth. While you are reading this 'some gigantic mass of magma that has been accumulating deep under the remote southern Andes may be priming itself to tear the crust apart – and our familiar world with it.' As McGuire himself has pointed out, his picture of our future is far from rosy. Tokyo is given the label 'the city waiting to die', and we will probably not have more than six months to prepare ourselves for a devastating meteorite impact.[23]

The fear and fascination of the biggest natural disasters contributes to removing focus from daily tragedies. Floods, drought, malnutrition, HIV/AIDS and malaria account for the greatest loss of human life today. So why has the fear of mega-disasters claimed so much attention? The reason may be that they do not have to do with a pressing reality but with a future scenario, and therefore border on fiction. So fascination and fear of these phenomena could be said to resemble the attitude of the eighteenth century towards nature. The philosopher Immanuel Kant's theory of fear-mingled joy in nature – the Sublime – presupposes that the observer is in safety. Apart from in a few countries, Europeans are, generally speaking, safe as regards natural disasters of epic dimensions.

Even so, we ought not to underestimate the dangers that surround us on the earth we exist on. Many of them we create ourselves, and they come from the borderland between nature and our societies. The threat assumed to be greatest the last decade comes from bird influenza, even though the hazard seems to have reduced in the last couple of years. Over 200 people in Asia have died of a form of influenza that has spread from birds to humans. A similar virus took the lives of between 20 and 100 million people as recently as 1918. If the virus mutates, the disease may be able to pass from human to human. The WHO estimated in 2005 that at worst 100 million people could die. Even according to more cautious estimates, several million could die. Others have claimed that as many as a billion people could lose their lives.[24] What distinguishes our fear of bird influenza from other comprehensive disasters is that it will also affect the rich parts of the world – at any rate, until a new vaccine appears.

CONSPIRACY THEORIES

Disasters cause alternative world pictures to emerge from its otherwise dark hiding place. An article on Al-Jazeera with the title 'The US Knew About the Tsunami', kept a lot of conspiracy-ridden heads busy for a long time after the tsunami disaster in 2004. 'Hakmed' from Iran claimed that 'this earthquake was caused by the us government to destroy the nation of indonesia. bush must be stopped so there can be peace in the world.' Others blamed Israel and Zionism. Similar debates went on in a number of Internet fora. 'Sahara 2' speculated on whether the CIA could be responsible. 'Isn't it even conceivable . . . that some form of explosive charge could be used, on one of the known 'fault lines' to induce an earthquake under the ocean? . . . I don't rule ANYTHING out, at this point'.[25] Even though the tsunami disaster was enormous, 'Hadeel' pointed out that 'we must also not forget the victem [sic] of the american weapons. more than 100,000 in iraq and most of them innocent civilians.'[26] Not all the discussions were equally fruitful, and 'Steebo' was depressed about those who blamed Allah. 'Come on man . . . I thought humanity had advanced beyond this primitive, ignorant mindset a couple of centuries ago.'[27]

The bluntest reaction to the tsunami disaster came from Westboro Baptist Church in Kansas. Three days after the disaster, the religious community issued a press release with the title 'Thank God for Tsunami & 2,000 Dead Swedes!!!'. They claimed the tsunami had been sent from God to take the lives of holidaying gay Swedes. The background for this reaction was the sentence passed on the Swedish Pentecostal Åke Green by a Swedish court, who had 'only' been guilty of condemning gays during a service. The Baptists in Kansas reacted by calling the Swedes 'homofascists' and Sweden the 'land of the sodomite damned'.[28] Green was acquitted during a new trial the following year.

It is difficult to know how seriously to take such moves, but they show the diversity of opinions and underground attitudes that can turn up after a disaster. The need to find scapegoats is strong, with many people including high political and profound religious conflicts in the fight to influence the understanding of reality.

Where the Devil Lives in the Ground

'The beginning is unfathomable. The end is unfathomable.'
Jewish Kabbalah

Darkness had descended on Timbuktu. We were on our way to Hachi Mohamed's house on the outskirts of the town, that clear evening in February. Only a few hours had passed since we had returned from the expedition to M'Bouna. After a short drive through the dusty streets of the town, we were met at Hachi's door.

The living rooms were painted white, the two windows covered by blue curtains. Along one of the walls was the sole item of furniture – a low bench with blue-patterned upholstery. With stiff knees after our fieldwork, we sat down on the large carpet in the middle of the floor. A pot of heavily sugared green tea was carried in and poured into small glasses. Pulses soon quickened.

Hachi talked about the civil war in the early 1990s and about the task of implementing the agreement between the authorities in Mali and the various groups of rebels. For several years, this part of the country had been subject to wars, violence and massacres, with Tuaregs, the army and local residents involved. Thousands of Tuaregs had fled to neighbouring countries. In 1995 peace talks finally got under way, and in spring 1996 an agreement was marked by a large weapons bonfire in Timbuktu. So far, 2,000 rebels had been integrated into society. That left 9,000, Hachi stated.

Then it was our turn to talk about the experiences we had had, about the meetings with the nomads, about the village of M'Bouna, and about our research. Our expedition had been successful, but the result totally unexpected.

Before we left M'Bouna we had planned a critical test. We wanted to dig for lava on the bed of the partially dried-out inland lake Lac Faguibine. The cars had stopped alongside a tongue-shaped depression, with hot fissures and a smoking front that was gradually advancing over the dry bed of the lake. The object of the test was to dig down as deep as possible right next to the smoking ground. After that, to dig inwards to find the source of heat. If there was lava there, it would fill up the hole.

Work began on the hole. The wind was strong and it was impossible to keep the sand out of one's eyes. The Mali geologists did not feel it was right for us to do the digging ourselves, so, after much palaver, they had brought in two men from M'Bouna to do the job. They took over. We stood there watching, even though we were itching to grab hold of the spades. Down in the hole it grew hotter and hotter. Was there a glow down at the bottom? We could not stand being onlookers any longer and took over. The roles were reversed, with neither the diggers nor the others able to conceal their mirth when they saw us at work. The spade dug through sand and clay, then through peat half a metre down. Smoke poured out from a fissure at the bottom. Suddenly, flames shot out of the layer of peat. The people around us shouted: 'Come and see. Come and see!'

The source of the high temperatures had been found. We went on digging. Had the fire been lit by the lava that lay directly beneath? In that case, it would soon fill the hole. But instead of finding glowing lava, all we found was cold sand under the peat. We stood staring into the hole, almost in disbelief. The new hypothesis we had been discussing the previous evening had been confirmed. The phenomenon had nothing to do with volcanic activity. We had discovered subsurface peat fires.

The fires had probably started in the same way as fires in bales of hay or in silos by heat being released when organic material is broken down. This meant that the residents of M'Bouna could feel safe. There were no volcanoes in the area, as the other researchers had claimed.[1]

Hachi was most interested in what we had to tell him. But he did not seem all that surprised when he heard about the fires. He himself had grown up in a village of nomads south of M'Bouna, close to an area called Daounas. Hachi explained that the heat and smoking ground was well known to the nomads. 'I learnt about it when I grew up. Neither travellers nor nomads spent the night in Daounas, because the devil was there. When you laid down at night, the flames came. So you always made sure of getting home first.' He also said that changes had taken place recently. It was not only single bushes that had caught fire but large areas. Complaints had come in from people in M'Bouna, and the problem had been discussed in the newspapers and on the radio. Hachi himself had seen the phenomenon at close quarters. 'It happened while I was attending secondary school. I was on my way to the neighbouring town with a friend and we spent the night outside. After darkness had fallen, we saw flames close by. We were so afraid that we quickly gathered our things together, mounted our camels and rode off. We had heard that the flames came from the devil.'

The devil lived in the ground, and the fires had become part of the local popular belief. Hachi's friend, who was sitting next to him, added that all those living in the area had had the same experiences: 'Today, many people think that there must be a scientific explanation for the phenomenon. But our local culture is based on the oral tradition. Very few people indeed can read and write, so attitudes and beliefs have persisted. Knowledge of the phenomenon has remained at the administrative level.'

We had deflated the disaster scenario that was brewing in Mali, but the phenomenon clearly led its own life and had done so for a long time. What more could be done? Hachi thought that when the local authorities heard about the new results, the people of the village would soon know about it. Word spreads quickly, especially among the nomads. But would they believe the conclusion reached? After all, not much time had passed since a group of researchers had expressed the view that there was a danger of a volcanic eruption. 'When the mayor says that it was these Europeans that have found out something, people will believe it. Then it is no longer the devil. Travellers will begin to sleep in Daounas again. The devil will disappear.'

The subsurface fires in Mali illustrate how potentially dangerous natural phenomena can be interwoven with societies and ideas. The story of the burning ground is also a narrative of adaptation. The starting point for the phenomenon was the drying out of the lake. During a drought in Sahel in the early twentieth century,

Frenchmen reported fires in the partially dried-out lake, but stated that they were put out when the water returned.

In Sahel the balance between the existence level of the population and fluctuations in precipitation is a delicate one, demonstrating the vulnerability of the society to even small climate variations. Over 100,000 people died during a long period of drought in the early 1970s.[2] A long time was to pass before the disaster was once more a fact. In 1984 thousands of people in Sahel starved to death. In Ethiopia, over a million died. The worst famine disaster in a century was a fact. We can perhaps best recall the images of the skinny African children that were shown on television.

People have lived in Sahel for thousands of years, even though the basis for existence here is vulnerable. The nomadic peoples were formerly able to move according to the climate variations in a way that is no longer possible. Several thousand years ago, the Sahara desert was actually a fertile place, with large lakes and a rich fauna. Hunters in the Stone Age were able to hunt both buffalo and elephants around the lakes at the time the pyramids in Egypt were being built. Our knowledge of the former climate of the Sahara comes from finds of fossils and cave paintings, which indicate that something dramatic took place about 4,500 years ago. An intense drought set in as a result of a global climate change, with only the Nile surviving as a constant source of water.[3]

But too much precipitation is also a threat. In autumn 2003 hundreds of flimsy clay houses in Timbuktu collapsed during heavy rains. Two children lost their lives. Paradoxically enough, large amounts of rain and floods are probably the only thing that can put out the subsurface fires.

The nomads in the areas around Timbuktu had a religious explanation for the fires – they were the work of the devil. But the belief that higher powers are behind natural disasters and natural phenomena is not limited to distant climes. Many of the examples of attitudes towards natural disasters in this book show the breadth of interpretations that are accessible, even in Western societies. Religious explanations of disasters have to do with people's worldview – not with superstition. Evangelistic Christian communities are estimated at having 550 million members worldwide, and most of the world's Christians now live outside Europe and North

America. There are about 1.3 billion Moslems, although it is uncertain how many of these are dedicated Moslems and follow the Koran. In addition, both Catholics and Jews make use of religious explanations of natural disasters. Perhaps as many as two billion people today include God in trying to explain natural disasters.

THE FUTURE OF NATURAL DISASTERS

The world community is facing enormous challenges to try and reduce the number of people who die from natural disasters. Fortunately, all went well in Sahel. No one in M'Bouna had to move because of the danger of a volcanic eruption. But many of the actual eruptions, floods, hurricane, earthquakes and periods of drought over the past 150 years have had vast consequences, and they all tell us something: something went very badly wrong. Let us learn from these mistakes so that they are not repeated.

After the publication of the IPCC reports in 2007, concerns grew about what climate changes can lead to. Pessimism increased. The combination 'IPCC' and 'doomsday' gave no less than 189,000 hits on Google in February 2007. The belief that 'the end is nigh' always flares up after major natural disasters. The new thing now is that climate changes projected forwards in time have the same effect. Climate changes are perceived by many as the most important scientific, political and societal problem of our yimes. The combination of rapid population increase, the formation of slums in urban areas and global climate changes will contribute to keeping the poor countries at the top of the disaster statistics.

It is thought-provoking that even with a reduction of gas emissions, and thereby a minimization of the assumed consequences of climate changes, about a hundred million people will still be affected by natural disasters, every single year. Taking a decade as our perspective, that could mean several hundred thousand people losing their lives. These are already hard-to-grasp figures that even without global warming and the IPCC's conclusions ought to have claimed worldwide media attention. With the history of natural disasters as a backdrop, today's situation must not be taken as the base line to measure future changes against. We are *not* living in an age without natural hazards and natural disasters. Measures to reduce the consequences of climate changes must therefore be more than a deal with emission quotas. The goal ought to be to achieve a real reduction of the number of people who are vulnerable to environmental changes. The key word is *adaptation*. Adaptation to climate changes was pushed into the background throughout the 1990s in relation to the reduction of emissions. The reason was that many people were afraid of a pragmatic attitude to climate changes. One example is Al Gore, who in 1992 claimed that adaptation was almost 'a kind of laziness, an arrogant faith in our ability to react in time to save our own skin.'[4] This attitude is in the process of changing. In the 1990s natural disasters were put on the agenda via the UN's international decade for natural disaster reduction (IDNDR). This programme has been continued under the banner 'International Strategy for Disaster Reduction', with the World Bank involved in some of the subprojects. The reduction of poverty has been given a central place in order to achieve the goal of reducing the extent of natural disasters. The roots of this approach,as we have seen, go back to the 1970s. But it will take a long while before the results of the the reduction of poverty will be visible. Meanwhile, new natural disasters cause that goal to slip even further away.

All the problems that can arise in a world with global warming are already common – and cause major problems. The reduction of greenhouse gas emissions will not be able to prevent disasters that are the result of social and economic distortion. Natural disasters such as Hurricane Katrina will arise whether we reduce the emissions of climate gases or not. Adaptation to future changes and natural hazards, whether they are the result of climate changes or not, will strengthen the ability of societies to withstand changes. If we succeed in this, it will represent a great victory.

MAJOR NATURAL DISASTERS

This overview is meant to provide examples of the enormous loss of human life that natural disasters have caused down through history. It is far from exhaustive, only including a selection of the biggest disasters. The sources are many (all are from publications or Internet addresses mentioned in the literature or in the notes), and the number of those who lost their lives often varies from source to source, so figures are approximate. Figures from official sources often list fewer casualties than, for example, those from independent international organizations. It is worth noting that even the simplified list adds up to more than 200 million victims.

AD 79	*Italy.* 3,300 died after the volcanic eruption of Vesuvius.
856	*Iran.* 200,000 died during an earthquake.
893	*Iran.* 150,000 died during an earthquake.
1138	*Syria.* 230,000 lost their lives during an earthquake.
1290	*China.* 100,000 lost their lives during an earthquake.
1348–9	*The Black Death strikes Europe.* About 50 million people lost their lives.
1556	*China.* 830,000 died after an earthquake and a number of landslides.
1586	*Indonesia.* 10,000 died during a volcanic eruption.
1642	*China.* Floods. More than 300,000 died.
1656	*Italy* (Naples). A plague epidemic claimed the lives of 300,000.
1667	*Caucasus.* Earthquake. 80,000 died.
1668	*Turkey.* 8,000 died after an earthquake in Anatolia.
1692–4	*France.* Famine and epidemics. 2 million died.
1693	*Italy.* Earthquake hit Sicily (54,000 died) and Naples (93,000 died).

1703	*Prussia and Lithuania.* Plague took the lives of 280,000.
1711	*Brandenburg.* Plague took the lives of 215,000.
1716	*Algeria.* Earthquake. 20,000 died.
1727	*Iran.* Earthquake. 77,000 died.
1737	*India.* Earthquake, tsunami and cyclone. Calcutta and Bengal hit, 300,000 died.
1740s	*Europe.* Famine and epidemics. Several hundred thousand died.
1755	*Portugal.* Up to 70,000 died as a result of an earthquake and a tsunami.
1759	*Syria.* Earthquake, 30,000 died. At the same time, 20,000 died in Lebanon.
1770	*India.* Famine and epidemics took the lives of 10 million people.
1770	*Moldavia.* 300,000 died of plague.
1780	*Iran.* 200,000 died after an earthquake.
1783	*Iceland.* 10,000 died as a result of famine and poisoning after the eruption of the Laki volcano. More than 45,000 died as a result of aftereffects elsewhere in Europe.
1783	*Italia.* Earthquake. 60,000 died.
1792	*Japan.* Volcanic eruption and tsunamis claimed the lives of 15,000.
1815	*Indonesia.* Volcanic eruption (Tambora), the most violent in recent times. Resulted in three years of global cooling. 92,000 died.
1857	*Italy.* An earthquake took the lives of 11,000 in Naples.
1876–9	*South America.* Between 1 and 2 million people died during a famine.
1870–78	*China.* Between 9 and 13 million people died during a famine.
1877–8	*India.* Famine. About 8 million people died.
1883	*Indonesia.* The volcanic eruption of Krakatau and a tsunami took the lives of 36,000.
1896–1900	*China.* Famine. 10 million people died.
1896–1902	*India.* Famine. Between 6 and 19 million people died.
1900	'The Galveston Hurricane'. 8,000 people died.
1902	*Martinique.* 30,000 died after a volcanic eruption (Mt Pelée) and pyroclastic flows.
1906	Earthquake that hit San Francisco claimed the lives of about 3,500 people.

1906	*Chile.* 20,000 died during an earthquake.
1908	*Italy.* 70–100,000 died during an earthquake in Messina.
1915	*Italy.* 30,000 died in an earthquake.
1918–19	*Large parts of the world.* Influenza pandemic. 50–100 million people died. Asia was the hardest hit, while 2.3 million people died in Europe. A total of 500 million people contracted the virus.
1920	*China.* 200,000 died during an earthquake.
1923	*Japan.* 142,800 died during an earthquake outside Tokyo.
1927	*China.* 200,000 died during an earthquake.
1928–30	*China.* Drought. 1.4 million died of hunger.
1931	*China.* 400,000 died during floods.
1932	*China.* 70,000 died during an earthquake.
1935	*China.* 142,000 died during floods.
1935	*Pakistan.* Earthquake. 30–60,000 died in the town of Quetta.
1941–2	*China.* Drought. 3 million died of hunger.
1943	*India* Famine. 3 million died.
1948	*Soviet Union (Turkmenistan).* 110,000 died during an earthquake.
1954	*China.* 30,000 died during floods.
1958–61	*China.* Famine related to 'The Great Leap Forward'. Between 20 and 40 million people died.
1970	*Bangladesh.* 300,000 people killed by cyclone.
1970	*Peru.* Earthquake and landslide claimed 67,000. The towns of Yungay and Ranrahirca were buried.
1971–3	*Sahel.* Drought. 100,000 died.
1972	*Nicaragua.* Earthquake. Between 5,000 and 20,000 died.
1974	*Honduras.* Hurricane Fifi, 8,000 died.
1976	*Guatemala.* 25,000 lost their lives during an earthquake.
1976	*China.* 250,000–800,000 died when the city Tangshan was flattened by an earthquake.
1978	*China.* Drought. 57,000 died.
1984	*Ethiopia.* Drought and famine. 1 million died.
1985	*Mexico.* Earthquake in Mexico City claimed the lives of 10,000.
1985	*Colombia.* Volcanic eruption of Nevado del Ruiz claimed the lives of 25,000.
1988	*Armenia.* Earthquake claimed the lives of 25,000.
1990	*Iran.* 40–50,000 died in an earthquake in Gilan province.

1991	*Bangladesh.* 130,000 died during a cyclone.
1993	*India.* 10,000 died during an earthquake.
1995	*Japan.* Earthquake hit the city of Kobe, claiming the lives of 6,400.
1997	*East Africa.* Floods and epidemics. 15,000 died.
1998	*Afghanistan.* Earthquake claimed the lives of 4,000.
1998	*Honduras.* Hurricane Mitch led to the deaths of about 11,000 people and to landslides and floods.
1998	*China.* Floods claimed the lives of 4,000 people. 300 million people were affected.
1999	*Turkey.* Earthquake claimed the lives of more than 17,000.
1999	*India.* Cyclone killed 10,000 and made 10 million people homeless.
2001	*India.* Earthquake killed almost 20,000 and made 1 million homeless.
2003	*Iran.* 26,000 killed during the earthquake that hit the city of Bam.
2003	*Europe.* Heat wave. 72,200 died (Italy, France and Spain accounted for 54,000 deaths).
2004	*South Asia.* Approximately 230,000 killed as a result of an earthquake and tsunami.
2005	Hurricane Katrina. 1,833 people died.
2005	*Pakistan.* Earthquake. Around 73,000 died and 3.5 million were made homeless.
2006	*Indonesia.* Earthquake. 5,780 died and 3.1 million people affected.
2006	*The Philippines.* Landslide in which 1,126 people lost their lives.
2006	*The Philippines.* Typhoon, 1,399 died.
2006	*Europe.* Heat wave. 3,300 died.
2007	*Bangladesh.* Cyclone Sidr. 4,200 died, 8.9 million people affected.
2007	*China.* Flood. 535 died, 105 million affected.
2007	*India.* Flood. 1,103 died, 18.7 million affected.
2008	*Burma.* Cyclone Nargis. At least 130,000 died.
2008	*China.* Earthquake. Between 70,000 and 88,000 died, and 5 million people made homeless.

REFERENCES

Preface: When Climate Becomes Disaster

1 'Annual Disaster Statistical Review: Numbers and trends 2006', at www.em-dat.net.
2 'Natural disasters more destructive than wars: Egeland', Yahoo! News, 28 August 2007.

Introduction: On the Edge of the Sahara

1 See T. A. Benjaminsen and G. Berge, *Timbuktu. Myter, mennesker, miljø* (Oslo, 2000) for an outline of the history of the Timbuktu area.
2 In addition we have the technological and social disasters. Technological hazards: linked to buildings, machinery, radio-activity, explosions, etc. Social hazards: weapons, poisonous gases, terrorists, armies (that can lead to wars, terrorism, genocide). See K. Hewitt, *Regions of Risk: A Geographical Introduction to Disasters* (Harrow, 1997). The most recent disaster classification was presented by www.em-dat.net in July 2008 (Cred Crunch number 13).

1 Mythologies

1 Sacrifice in the pre-Christian era is the theme of B.-M. Näsström's book *Blot. Tro og offer i det førkristne Norden* (Oslo, 2001).
2 Translation by Benjamin Thorpe at http://www.northvegr.org/lore/poetic2/001_02.php.
3 See for example P. A. Munch, *Norrøne gude- og heltesagn* (Universitetsforlaget, 1996), for more material about Ragnarok and Norse mythology.

4 See J. Hanska, 'Strategies of Sanity and Survival: Religious Responses to Natural Disasters in the Middle Ages', *Studia Fennica Historica*, ii (Helsinki, 2002).

5 The two quotes in this section are from Ø. Hodne, *Jutulhugg og riddersprang. Sagn fra norsk natur* (Oslo, 1990).

6 Näsström, *Blot*.

7 See A. B. Amundsen, 'Konventikler og vekkelser', in *Norges religionshistorie*, ed. Bugge Amundsen (Universitetsforlaget, 2005), pp. 295–316.

8 Most of the information about natural disasters in the Middle Ages is from Hanska, 'Strategies of Sanity and Survival', unless otherwise specified.

9 This view is supported by Hanska, 'Strategies of Sanity and Survival'.

10 See O. J. Benedictow, *The Black Death 1346–1353: The Complete History* (Woodbridge, 2004), for an updated account of mortality during The Black Death.

11 Näsström, *Blot*.

12 See for example E. L. Quarantelli, ed., *What is a Disaster? Perspectives on the Question* (London and New York, 1998) for a thorough discussion of the classification of disasters and, as mentioned above, www.em-dat.net for a recent update (Cred Crunch 13).

13 N. Cohn, *The Pursuit of the Millennium: Revolutionary Millenarians and Mystical Anarchists of the Middle Ages* (London, 1993), p. 127.

14 Ibid., p. 136.

15 H. E. Næss, *Trolldomsprosessene i Norge på 1500–1600 tallet* (Universitetsforlaget, 1982).

16 The study is by the French historian Jean Delumeau, as mentioned in Hanska, 'Strategies of Sanity and Survival'.

17 Benedictow, *The Black Death 1346–1353*, R. S. Gottfried, *The Black Death: Natural and Human Disasters in Medieval Europe* (New York 1983).

18 N. Gilje, and T. Rasmussen, *Norsk Idéhistorie*, vol. ii: *Tankeliv i den lutherske stat* (Oslo, 2002).

19 Cohn (1999).

20 C. Merchant, *Reinventing Eden: The Fate of Nature in Western Culture* (New York and London, 2004).

21 Material about the earthquake in London in 1750 is from T. D. Kendrick, *The Lisbon Earthquake* (Philadelphia and New York, 1957).

22 From www.hammondindiana.com/20thcentury/timecapmain.htm.

23 *The New York Times*, Tuesday, 15 April 2008.

24 From www.hammondindiana.com/20thcentury/timecapmain.htm.

25 B. Hedberg, *Kometer och kometskräck* 1985.

26 Ø. Hodne, *Norsk folketro* (Oslo, 1999).

2 The Day of the Dead

1 J. L. Justo and C. Salwa, 'The 1531 Lisbon Earthquake', *Bulletin of the Seismological Society of America*, LXXXVIII (1998), pp. 319–28.

2 It is estimated that the strength of the quake corresponded to 8.5 on the Richter scale (see D. Chester, 'The 1755 Lisbon Earthquake', Progress in Physical Geography, XXV (2001), pp. 363–83). This estimate is based on the intensity of the earthquake. Intensity is measured on the so-called Modified Mercalli scale and is based on the effect on people, buildings and the ground. The scale is from I to XII and is thus descriptive. The intensity of an earthquake can, then, be reconstructed from eyewitness accounts and other historical information. This can subsequently be converted into the Richter scale if so desired. Both scales are logarithmic, that is, each individual increase in the scale corresponds to a tenfold increase in earthquake strength.

3 The literature about the Lisbon disaster is comprehensive and the amount of source material large. Unless otherwise specified, the historical sources for this chapter have mainly been T. D. Kendrick, *The Lisbon Earthquake* (Philadelphia and New York, 1957); D. Birmingham, *A Concise History of Portugal* (Cambridge, 1993); R. R. Dynes, 'The Dialogue Between Voltaire and Rousseau on the Lisbon Earthquake: The Emergence of a Social Science View', *International Journal of Mass Emergencies and Disasters*, XVIII (2000), pp. 97–115; Chester, 'The 1755 Lisbon Earthquake', and J. D. Fonseca, *1755: The Lisbon Earthquake* (Lisbon, 2005). The sources for the size and extent of the earthquake, as well as the geology of the area, are based on Chester, 'The 1755 Lisbon Earthquake', and M. A. Baptista et al., 'New Study of the 1755 Earthquake Source Based on Multi-channel Seismic Survey Data and Tsunami Modeling', *Natural Hazards and Earth System Sciences*, III (2003), pp. 333–40.

4 A. Dyregrov, *Disaster Psychology* (Bergen, 2002).

5 Terra Nova, 'Lisbon Recalled: All Saints Day, 1 November 1755',

Terra Nova, III (1991), pp. 670–72. The account was written down for The British Historical Society of Lisbon, and is from 1755.

6 M. Carozzi, 'Reaction of British Colonies in America to the 1755 Lisbon Earthquake', *History of Geology*, II (1983), pp. 17–27.

7 Birmingham, *A Concise History of Portugal*.

8 A. H. De Oliveira Marques, *History of Portugal*, vol. I, *From Lusitana to Empire* (New York and London, 1972).

9 http://www.literature.org/authors/voltaire/candide/chapter-06.html.

10 H. Strøm, *Physisk og Oeconomisk Beskrivelse over Fogderiet Søndmør, beliggende i Bergens Stift i Norge* (Kiøbenhavn, 1762–6).

11 Observation from Randsfjorden, southwest of Oslo in Norway.

12 *Kiøbenhavnske Danske Post-Tidende* was established in 1749 by the printer Ernst Henrich Berling and is considered to be Denmark's first newspaper. It came out twice a week (Monday and Friday) and had its own advertisement supplement. It later came to be called *Berlingske Tidende* – and it still exists.

13 Carozzi, 'Reaction of British Colonies in America to the 1755 Lisbon Earthquake'.

14 Among those who took part was the Danish theologian Hans Adolph Brorson (1694–1764), who wrote a rhyming poem of no less than 380 lines with the title 'The Piteous Ruin and Destruction of Lisbon I/II'. Like many others, Brorson had read about the disaster in *Kiøbenhavnske Danske Post-Tidende*. He build up a mood of Lisbon as heaven on earth – a city that was admired by all of Europe. Similar assessments of Lisbon flourished after the disaster, even though it had previously been claimed that it was both dirty and ugly. Brorson was typical of his age in believing that the quake was anchored in physical nature but had to be seen as a sign of God's wrath, and the mankind was heading for the latter days and the final apocalypse. Brorson is considered one of Denmark's most important Pietists of the eighteenth century. More information about Brorson and his works can be found in Arkiv for Dansk Litteratur, at www.adl.dk.

15 Prayer days in Denmark and on Iceland are mentioned in the diary of the Icelandic clergyman Jón Steingrímsson, Fires of the Earth (Rejkjavik, 1998) from the 1780s.

16 The Danish title of his book is 'Uforgribelige betænkninger over den naturlige aarsag til de mange og store jordskiælv, samt det

usædvanlige veirlig, som nu paa nogen tid er fornummet, baade i
og uden for Europa'. It was published in Copenhagen in 1756.

17 Biographical information about Pontoppidan has mainly been
taken from Dansk Biografisk Leksikon. The theological aspects of
Pontoppidan's work are dealt with in N. Gilje and T. Rasmussen,
Norsk Idéhistorie, vol. II: *Tankeliv i den lutherske stat* (Oslo, 2002).

18 'Det første Forsøg paa Norges Naturlige Historie' was published in
two volumes (1752–3) and was soon translated into German and
English.

19 In Europe there were many who supported the idea that God had
something to do with the disaster. The traces of these are, on the
other hand, indistinct in Denmark–Norway. We must recall that it
was mainly the philosophers of the Enlightenment and certain
natural scientists that wished to keep God out of the discussion. In
Norway there was not even a university in the eighteenth century,
while the natural sciences at the university in Copenhagen were
still dominated by the clergy. So there was little fertile ground for
independent comments, with most debates being of a theological
nature. What people in general felt about the disaster is not known
for certain, but we can assume that all of those who heard about it
reacted with disbelief and dread.

20 For a thorough account of creationism, see R. T. Pennock, *Tower
of Babel: The Evidence Against the New Creationism* (Cambridge, MA,
2002).

21 Opinion polls carried out in 1995 show a steady high level of about
45 per cent who answered yes to the statement 'God Created Man
in his Image 10,000 years Ago', *New Scientist*, 9 July 2005.

22 'Britons Unconvinced on Evolution', BBC website, 26 January 2006.

3 California: Earthquake and Culture

1 At www.fordham.edu/halsall/basis/omarkhayyam-rub2.html.

2 Daily updates and maps of earthquakes in southern California can
be found at the USGS website, www.usgs.gov.

3 See H. Svensen et al., 'Processes Controlling Water and
Hydrocarbon Composition in Seeps from the Salton Sea
Geothermal System, California, USA', *Geology*, XXXV (2007), pp. 85–8,
for a detailed account of the hot springs.

4 The brochure can be downloaded from www.scec.org.

5 The film *Volcano* is dealt with by S. Keane in the book *Disaster Movies: The Cinema of Catastrophe* (London and New York, 2001).

6 US Geological Survey, at www.usgs.gov.

7 M. Davis, *Ecology of Fear: Los Angeles and the Imagination of Disaster* (Basingstoke and Oxford, 1998).

8 Ibid.

9 T. Steinberg, 'Smoke and Mirrors: The San Francisco Earthquake and Seismic Denial', in *American Disasters*, ed. Steven Biel (New York, 2001), pp. 103–28.

10 Steinberg in ibid. mentions that a leading article in the newspaper *San Francisco Call*, owned by the capitalist John Spreckels, protested, expressing the opinion that all the publicity surround the disaster would scare people out of the city. They wanted the disaster to be forgotten as soon as possible.

11 Ibid.

12 Davis, *Ecology of Fear*.

13 The story was printed in *The New York Times*, 12 July 1936, see Steinberg, 'Smoke and Mirrors'.

14 The story has been taken from C. Morris, *The San Francisco Calamity by Earthquake and Fire* (Urbana, IL, 2002).

15 The text can be found on the USGS site about the disaster: www.usgs.gov.

16 Joaquin Miller wrote about his experiences for Oakland Tribune, 6 May 1906. The text can be found at the USGS website about the disaster: www.usgs.gov.

17 www.sfmuseum.org/hist5/jlondon.html.

18 Extracts from 'The Story of an Eyewitness', from the website of San Francisco city museums: www.sfmuseum.net.

19 R. H. Platt, *Disasters and Democracy: The Politics of Extreme Natural Events* (Washington, DC, 1999).

20 http://pubs.usgs.gov/fs/1999/fs152-99.

21 Information about the meeting and content of the article from the *Los Angeles Daily Times* has been taken from the website www.dunamai.com.

22 Frank Bartleman (1871–1936) wrote more than 550 articles, 100 tracts and 6 books, according to the epilogue of his book *Azusa Street* (New Kensington, PA, 1982).

23 Ibid., p. 38.

24 Ibid., p. 49.

25 I was granted access to 'The Earthquake!!!' by The Methodist Heritage Center in the USA.

26 A. Anderson, *An Introduction to Pentecostalism* (Cambridge, 2004).

27 M. Davis, *Dead Cities* (New York, 2002).

28 Information about the after-effects of the earthquake in Charleston are from T. Steinberg, *Acts of God: The Unnatural History of Natural Disasters in America* (Oxford, 2000).

29 From the Gospel according to St Mark.

4 Natural Disasters in Metropolis

1 B. Chen, '"Resist the Earthquake and Rescue Ourselves": The Reconstruction of Tangshan after the 1976 Earthquake', 1, *The Resilient City: How Modern Cities Recover from Disasters*, ed. L. J. Vale and T. J. Campanella (Oxford, 2005), pp. 235–53.

2 Ibid.

3 'China Eases State Secrets Control', BBC website, 12 September 2005.

4 E. Galeano, *Memory of Fire*, III, *Century of the Wind* (New York, 1998), p. 211.

5 See J. Zeilinga de Boer and D. T. Sanders, *Earthquakes in Human History: The Far-reaching Effects of Seismic Disruptions* (Princeton, NJ, and Oxford, 2005) for more information about the earthquake and its role in the history of Nicaragua.

6 R. W. Kates et al., 'Human Impact of the Managua Earthquake', *Science*, CLXXXII, 7 December 1973, pp. 981–90.

7 G. Black, *Triumph of the People: The Sandinista Revolution in Nicaragua* (London, 1981).

8 Galeano, *Memory of Fire*, III, *Century of the Wind*, p. 211.

9 J. M. Albala-Bertrand, *Political Economy of Large Natural Disasters* (Oxford, 1993). Black, *Triumph of the People*.

10 M. Pelling, *The Vulnerability of Cities: Natural Disasters and Social Resilience* (London and Sterling, VA, 2003).

11 A. Oliver-Smith, 'Peru's Five Hundred-Year Earthquake: Vulnerability in Historical Context', in *The Angry Earth: Disaster in Anthropological Perspective*, ed. A. Oliver-Smith and S. M. Hoffman (New York and London, 1999), pp. 74–88.

12 For more information about the increase of disasters over the past 50 years, see the International Disasters Database at www.em-dat.net.

13 Quoted in Pelling, *The Vulnerability of Cities*, p. 27.

14 The figure applies for the period 1974–94. K. Hewitt, *Regions of Risk: A Geographical Introduction to Disasters* (Harrow, 1997).

15 Pelling, *The Vulnerability of Cities*.

16 Hewitt, *Regions of Risk*.

17 See Fothergill et al., 'Race, Ethnicity and Disasters in the United States: A Review of the Literature', *Disaster*, XXIII (1999), pp. 156–73.

18 The study is mentioned in ibid.

19 See for example D. Brinkley, *The Great Deluge: Hurricane Katrina, New Orleans, and the Mississippi Gulf Coast* (New York, 2006) for a detailed account of the course of the disaster.

20 From a press conference at New Orleans International Airport. www.whitehouse.gov/news/releases/2005/09/.

21 M. Davis, 'The Predators of New Orleans', *Le Monde diplomatique*, October 2005.

22 The poverty line in the USA corresponds to an income of $14,680 per year for a family of three. *Newsweek*, 19 September 2005.

23 *Time*, 19 September 2005.

24 Davis, 'The Predators of New Orleans'.

25 Ibid.

26 'New Orleans Lays Off 3,000 Workers', *Guardian*, 5 October 2005.

27 Davis, 'The Predators of New Orleans'.

28 'New Orleans' Homeless Rate Swells to 1 in 25', *USA Today*, 16 March 2008.

29 'World Press: Katrina "Testing us"', BBC website, 5 September 2005.

30 'Katrina Showed us Could be Turned Into "War Zone"', aljazeera.com, 12 September 2005. The researcher Ilan Kelman claims that Iran did not behave particularly aggressively after the hurricane and that the country actually accepted the USA's offer of aid after the earthquake disaster in BAM in 2003. I. Kelman, 'Hurricane Katrina Disaster Diplomacy', *Disasters*, XXXI, pp. 288–309.

31 'World Press: Katrina "Testing us"'.

32 See Kelman, 'Hurricane Katrina Disaster Diplomacy'.

33 From a press conference held by Farrakhan in Virginia, 3 September 2005. www.finalcall.com.

34 www.christianlifeandliberty.net/.

35 D. Godrej, *The No-Nonsense Guide to Climate Change* (London, 2001).

36 A. K. Geonjian et al., 'Posttraumatic Stress and Depressive Reactions Among Nicaraguan Adolescents after Hurricane Mitch',

American Journal of Psychiatry, CLVIII (2001), pp. 788–94.

37 Davis, 'The Predators of New Orleans'.

38 M. Davis, *Dead Cities* (New York, 2002). See
http://sunsite.berkeley.edu/ for a collection of Jack London texts
on the Internet.

39 D. Alexander, 'Symbolic and Practical Interpretations of the
Hurricane Katrina Disaster in New Orleans', at
http://understandingkatrina.ssrc.org (2005).

40 L. J. Vale and T. J. Campanella, 'The Cities Rise Again', in *The
Resilient City: How Modern Cities Recover from Disasters*, ed. T. J. Vale
and L. J. Campanella (Oxford, 2005), pp. 3–23.

41 L. J. Vale and T. J. Campanella, 'Axioms of Resilience', in *The
Resilient City*, pp. 335–55.

42 This approach to vulnerability is adopted from Pelling, *The
Vulnerability of Cities*, but the term is used in other ways as well. See
Wisner et al., *At Risk: Natural Hazards, People's Vulnerability and
Disasters* (London and New York, 2004), pp. 13–20, and Hewitt,
Regions of Risk, pp. 141–68, for a broader discussion.

43 Hewitt, *Regions of Risk*.

44 J. Borland, 'Stories of Ideal Japanese Subjects from the Great
Kantō Earthquake of 1923', *Japanese Studies*, XXV (2005), pp. 21–34.

45 Mentioned in a radio address, 10 September 2005:
www.whitehouse.gov/news/releases/2005/09/20050910.html.

5 Among the High Mountains and Deep Fjords

1 Figures from www.snoskred.no.

2 A. Furseth, *Dommedagsfjellet. Tafjord 1934* (Gyldendal, 1994).

3 For more about the Storega slide, see for example S. Bondevik et
al., 'The Storegga Tsunami along the Norwegian Coast, its Age
and Runup', *Boreas*, XXVI (1997), pp. 29–53.

4 The information is from www.skrednett.no.

5 A number of researchers from the Geological Survey of Norway,
Norges geotekniske institutt and the University of Oslo are
involved in the project.

6 Furseth, *Dommedagsfjellet*.

7 Ibid.

8 Based on ibid.

9 Ibid.

10 *The Times*, Monday, 9 April 1934.

11 *Aftenposten*, 10 April 1934 (leader).

12 The section is based on Furseth, *Dommedagsfjellet.*

13 At www.ngu.no.

14 From a conversation with Rev. Runde in April 2005.

15 Information about Ibsen's stay in Hellesylt is from www.hellesylt.no and www.ibsen.net.

16 O. Barman, *Erindringer fra 1861 til 1867* (Trondhjem, 1904).

17 H. Strøm, *Physisk og Oeconomisk Beskrivelse over Fogderiet Søndmør, beliggende i Bergens Stift i Norge* (Kiøbenhavn, 1762–6), vol. I.

18 N. Gilje and T. Rasmussen, *Norsk Idéhistorie*, vol. II: *Tankeliv i den lutherske stat* (Oslo, 2002).

19 Ibsen's view of nature is discussed in O. R. Søberg, 'En analyse av naturmotivene i Henrik Ibsens "Brand"', Masters thesis, University of Oslo (1995).

20 Extract from the poem 'På viddene' by Henrik Ibsen from 1859.

21 The story of the 'sight' was told me by Anders Bødal in a conversation with him in April 2005.

22 G. Svendsen and K. Werswick, *Fjellene dreper. Utsnitt av norsk katastrofehistorie gjennom 300 år* (Oslo, 1961).

23 Ibid.

24 S. Nesdal, *Lodalen – fager og fårleg* (Oslo, 2003).

25 Svendsen and Werswick, *Fjellene dreper.*

26 Ibid.

27 Recollections of Rasmus Nesdal, reproduced in Nesdal, *Lodalen.*

28 *Aftenposten*, 16 and 17 September 1936.

29 Quotation from an Oslo newspaper after the disaster in 1936. From Svendsen and Werswick, *Fjellene dreper.*

30 *Aftenposten*, 15 September 1936, p. 8.

31 A. Nesje, 'Ikkje gløymt etter 100 år.' GEO, IV (2005), pp. 34–8.

32 A. Dyregrov, *Disaster Psychology* (Bergen, 2002).

33 See D. Alexander, *Natural Disasters* (New York and London, 2001), for an outline of the geological relations that caused the landslide.

34 A. Favaro et al., 'Surviving the Vajont Disaster: Psychiatric Consequences 36 Years Later', *The Journal of Nervous and Mental Disease*, CXCII (2004), pp. 227–31.

35 See Dyregrov, *Disaster Psychology*, for more information about PTSD.

36 The statement is from a man in Oldedalen, 1996. From T. Larsen, 'Ras, risiko og rasjonalitet – en studie av et skredtruet samfunn på

Vestlandet', Masters thesis, University of Bergen (1998).

37 The database is a survey with 133 people asked, ibid.

38 Ibid.

39 Ø. Støylen, 'Å leve med skredfare. En studie om opplevelse av risiko og psykologiske belastninger', Masters thesis, University of Bergen (1999).

40 Furseth, *Dommedagsfjellet*.

41 From a conversation with Bødal in April 2005.

42 Referred to in G. Ingebrigtsen et al., 'Når ulykken rammer. Posttraumatisk stressforstyrrelse i Norge, forekomst og relasjon til sosialt nettverk', i: *Sosialt nettverk, helse og samfunn*, ed. by O. Dalgard, E. Døhlie & M. Ystgaard (Universitetsforlaget, 1993), pp. 64–85.

43 O. Barman, *Erindringer fra 1861 til 1867* (Trondhjem, 1904).

44 A. Møller, *Den farlege naturen* (Oslo, 1986), p. 199.

45 D. Alexander, *Natural Disasters* (New York and London, 2001).

46 W. Burroughs, ed., *Climate into the 21st Century* (Cambridge, 2003).

47 P. L. Doughty, 'Plan and Pattern in Reaction to Earthquake: Peru, 1970–1998', in *The Angry Earth: Disaster in Anthropological Perspective*, ed. A. Oliver-Smith and S. L. Hoffman, pp. 234–56 (New York and London, 1999).

48 N. Fulsås, *Havet, døden og vêret. Kulturell modernisering i Kyst-Noreg 1850–1950* (Oslo, 2003).

49 This section is based on ibid.

50 The information about the protests concerning swimming lessons is from Tore Frost (Associate Professor of Philosophy at Oslo University).

51 Matthew 24:7.

52 Information about the Man from Lebesby and his prophecies have been taken from the Internet magazine 'the prophetic voice' (www.profetier.com), written by Gunnar Lund.

53 T. Gunnarson, *Dommedagsventing. Millennismen og dens innslag i norsk kristendom* (Bergen, 1928).

6 Problem Children

1 E. Galeano, *Memory of Fire*, III, *Century of the Wind* (New York, 1998), pp. 6–7.

2 Ibid., p. 7.

3 D. Wilkinson, *Silence on the Mountain: Stories of Terror, Betrayal, and*

Forgetting in Guatemala (Boston and New York, 2001).

4 Geological information is based on S. N. Williams and S. Self, 'The October 1902 Plinian Eruption of Santa Maria Volcano, Guatemala', *Journal of Volcanology and Geothermal Research*, XVI (1983), pp. 33–56.

5 H. Sigurdsson and S. Carey, 'Volcanic Disasters in Latin America and the 13th November 1985 Eruption of Nevado del Ruiz Volcano in Colombia', *Disasters*, X (1986), pp. 205–16.

6 The story of the farmer is from www.paricutin.com and www.geology.sdsu.edu/how_volcanoes_work/Paricutin.html (where the quote is taken from).

7 Based on M. L. Nolan, 'Impact of Parícutin on Five Communities', in *Volcanic Activity and Human Ecology*, ed. P. Sheets and D. Grayson (Boston, MA, 1979), pp. 293–338.

8 D. Chester, *Volcanoes and Society* (London, Melbourne and Auckland, 1993).

9 Sigurdsson and Carey, 'Volcanic Disasters in Latin America and the 13th November 1985 Eruption of Nevado del Ruiz Volcano in Colombia'.

10 V. Bruce, *No Apparent Danger: The True Story of Volcanic Disaster at Galeras and Nevado del Ruiz* (New York, 2001), p. 19.

11 Ibid, p. 64.

12 The history of Nevado Del Ruiz is based on Bruce, *No Apparent Danger*; B. Voight, 'The 1985 Nevado del Ruiz Volcano Catastrophe: Anatomy and Retrospection', *Journal of Volcanology and Geothermal Research*, XLIV (1990), pp. 349–86, and Sigurdsson and Carey, 'Volcanic Disasters in Latin America and the 13th November 1985 Eruption of Nevado del Ruiz volcano in Colombia'. The first two in particular give a detailed account of the events.

13 Voight, 'The 1985 Nevado del Ruiz Volcano Catastrophe', p. 383.

14 Ibid, p. 383.

15 R. J. Blong, *Volcanic Hazards. A Sourcebook on the Effects of Eruptions* (Boston, MA, 1984).

16 J. Steingrímsson, *Fires of the Earth* [1907] (Reykjavik, 1998), p. 25.

17 Ibid, p. 27.

18 Ibid, p. 77.

19 Ibid, p. 78.

20 Ibid, p. 64.

21 Ibid, p. 88.

22 Ibid, p. 89.

23 R. Stone, 'Iceland's Doomsday Scenario?', *Science*, CCCVI (2004), pp. 1278–81.

24 J. Grattan, 'Pollution and Paradigms: Lessons from Icelandic Volcanism for Continental Flood Basalt Studies', *Lithos*, LXXIX (2005), pp. 343–53.

25 Stone, 'Iceland's Doomsday Scenario?'.

26 S. Thorarinsson, 'Damage Caused by Volcanic Eruptions', I: *Volcanic Activity and Human Ecology*, ed. Sheets and Grayson, pp. 125–59.

27 B. Fagan, *The Little Ice Age: How Climate Made History 1300–1850* (New York, 2002). See also J. Zeilinga de Boer and D. T. Sanders, *Earthquakes in Human History: The Far-reaching Effects of Seismic Disruptions* (Princeton, NJ, and Oxford, 2005) for a good account of the eruption in 1815 and the global consequences.

28 Á. Gunnarsson, *Volcano: Ordeal by Fire in Iceland's Westmann Islands* (Rejkjavik, 1973), p. 48.

29 R. S. Williams and J. G. Moore, 'Man Against Volcano. The Eruption of Heimaey, Vestmannaeyjer, Iceland', at www.ugs.gov (1983).

30 Chester, *Volcanoes and Society*.

31 See for example K. Hewitt, *Regions of Risk: A Geographical Introduction to Disasters* (Harrow, 1997).

32 E. Stangeland, 'Å leve på en vulkan. Risiko og stedsidentitet på Vestmannaeyjar', Masters thesis, University of Oslo (2004).

33 The survey is referred to in Chester, *Volcanoes and Society* but originally comes from a work by Clapperton from 1973.

34 Stangeland, 'Å leve på en vulkan'.

35 M. Davis, *Ecology of Fear: Los Angeles and the Imagination of Disaster* (Basingstoke and Oxford, 1998).

36 Stangeland, 'Å leve på en vulkan'.

37 Ibid.

38 Ibid.

39 P. R. Adams and G. R. Adams, 'Mount Saint Helens's Ashfall: Evidence for a Disaster Stress Reaction', *American Psychologist*, XXXIX (1984), pp. 252–60.

40 S. H. Ambrose, 'Late Pleistocene Human Population Bottlenecks, Volcanic Winter, and Differentiation of Modern Humans', *Journal of Human Evolution*, XXXIV (1998), pp. 623–51.

41 T. Hallam, *Catastrophes and Lesser Calamities: The Causes of Mass Extinctions* (Oxford, 2004).

42 K. A. Jacobsen, *Hinduism* (Oslo, 2003).

43 D. Cyranoski, 'A Sleeping Giant Stirs', *Nature*, CDXXVIII (2004), pp. 12–13.

44 See www.sacredsites.com/asia/japan, and J. Herbert, *Shintô: At the Fountain-Head of Japan* (London, 1967).

7 The Politics of Disasters

1 Albert Camus, *The Plague*, p. 284.

2 The weather data in this paragraph are from P. D. Jones and K. R. Briffa, 'Unusual Climate in Northwest Europe during the Period 1730 to 1745 Based on Instrumental and Documentary Data', Climate Change, LXXIX (2006), pp. 361–79.

3 J. D. Post, *Food Shortage, Climate Variability, and Epidemic Disease in Preindustrial Europe: The Mortality Peak in the Early 1740s* (Ithaca, NY, and London, 1985), p. 246.

4 Ibid.

5 J. D. Post, 'Nutritional Status andMortality in Eighteenth-Century Europe', in *Hunger in History: Food Shortage, Poverty and Deprivation*, ed. L. F. Newman (Cambridge, MA, 1990) (1990), p. 245.

6 B. Fagan, *The Little Ice Age: How Climate Made History 1300–1850* (New York, 2002), p. 141.

7 Post, *Food Shortage, Climate Variability, and Epidemic Disease in Preindustrial Europe*.

8 J. Herstad, *I helstatens grep. Kornmonopolet 1735–88* (Oslo, 2000).

9 Fagan, *The Little Ice Age*.

10 Psot, *Food Shrotage, Climate Variability, and Epidemic Disease in Preindustrial Europe*.

11 J. Herstad, *I helstatens grep. Kornmonopolet 1735–88* (Oslo, 2000), and Post, *Food Shortage, Climate Variability, and Epidemic Disease in Preindustrial Europe*. The corn market in Denmark and Norway was protected in the eighteenth century by restrictive legislation and tariff walls. In order to ensure high prices for grain and a high income for Denmark a total ban on the import of grain to Norway from other countries than Denmark was introduced in 1735. As the famine grew worse in summer 1741, the import ban on grain was raised as an emergency measure as well as a temporary ban on

exporting grain *from* Norway.

12 Jones and Briffa (2006).

13 Briffa et al. (1998).

14 Something happened to the climate in North America around 1740 too. Data from tree rings from California and Oregon show that the growth zone corresponding to 1739 was the thinnest since the beginning of records (corresponding to 1420). See 'Climatic Assessment of a 580-year *Chamaecyparis Lawsonia* (Port Oxford Cedar) Tree-ring Chronology in the Siskiyou Mountains, USA', by Carroll and Jules in the journal Madroño, LII (2005), pp. 114-122.

15 See Post, *Food Shortage, Climate Variability, and Epidemic Disease in Preindustrial Europe,* for further information about the crisis in the 1770s.

16 Most of the chapter is based on M. Davis, *Late Victorian Holocausts: El Niño Famines and the Making of the Third World* (London and New York, 2002).

17 Davis, *Late Victorian Holocausts,* p. 48.

18 Ibid., p. 58.

19 Ibid., p. 9.

20 Ibid.

21 Quoted in ibid., p. 76, from a nineteenth-century Chinese source.

22 D. Arnold, 'Hunger in the Garden of Plenty: The Bengal Famine of 1770', in *Dreadful Visitations. Confronting Natural Catastrophe in the Age of Enlightenment,* ed. Alessa Johns (New York and London, 1999), p. 96.

23 Ibid.

24 Davis, *Late Victorian Holocausts,* p. 139.

25 See for example Christian Eckert, 'The Next Tsunami Coming to Sri Lanka Will Be a Religious One', www.buddhistchannel.tv, 22 May 2005. Similar stories abounded, from both Sri Lanka and Indonesia, but they are often badly documented.

26 Fagan, *The Little Ice Age.*

27 IPCC 2001, *Climate Change 2001: The Scientific Basis* (Cambridge, 2001).

28 R. W. Davies and S. G. Wheatcroft, *The Years of Hunger: Soviet Agriculture, 1931–1933* (Basingstoke, 2004).

29 Davis, *Late Victorian Holocausts.*

30 M. Pelling, *The Vulnerability of Cities: Natural Disasters and Social Resilience* (London and Sterling, VA, 2003).

31 B. Wisner et al., *At Risk: Natural Hazards, People's Vulnerability and*

Disasters (London and New York, 2004).

32 A. K. Sen, *Poverty and Famines: An Essay on Entitlement and Deprivation* (Oxford, 1981).

8 Climatic Disasters

1 W. F. Ruddiman 'The Anthropogenic Greenhouse Era Began Thousands of Years Ago', *Climatic Change*, LXI (2003), pp. 261–93.

2 If one goes back to prehistoric times, our major cultural changes often coincided with climate changes. See for example J. Diamond, *Collapse: How Societies Choose to Fail or Survive* (London and New York, 2005) and S. Haberle and A. Lusty, 'Can Climate Influence Cultural Development? A View through Time', *Environment and History*, VI, pp. 349–69.

3 An outline of the history of research into climate change in the twentieth century is provided by S. R. Weart, *The Discovery of Global Warming* (Cambridge and London, 2003).

4 Ibid., p. 145.

5 Quoted in ibid., page 172.

6 IPCC 2001, *Climate Change 2001: The Scientific Basis* (Cambridge, 2001), 'Synthesis Report, Summary for Policymakers', p. 5. For a fascinating account of the political tussles surrounding the IPCC reports, see Jeremy Leggett's book *The Carbon War* (1999).

7 IPCC 2007, *Climate Change 2007*, at www.ipcc.ch. See for example 'Summary for Policymakers'.

8 All figures are from www.em-dat.net.

9 IPCC 2001, *Climate Change 2001*.

10 D. Godrej, *The No-Nonsense Guide to Climate Change* (London, 2001). The estimate applies for 2001.

11 www.who.org.

12 www.em-dat.be, *Cred Crunch* newsletter, issue no. 9. There are great difficulties in estimating the number of dead during a heat wave, and the figure presented here is a statistical estimate. For other parts of the world than Europe little is still known about the number of people dead as a result of extreme temperatures.

13 The figures are from www.em-dat.be, with a reservation for an under-reporting of natural disasters, especially heat waves prior to 1970.

9 The Tsunami

1 Quote from the book *Signs of the Last Day*, p. 52, downloaded at www.harunyahya.com.

2 S. Winchester, *Krakatoa: The Day the World Exploded, 27 August 1883* (London and New York, 2004).

3 R. Bilham, 'A Flying Start, Then a Slow Slip', *Science*, CCCVIII (2005), pp. 1126–17, and T. Lay et al., 'The Great Sumatra-Andaman Earthquake of 26 December 2004', *Science*, CCCVIII (2005), pp. 1127–32 (2005).

4 R. S. Stein, 'Earthquake Conversations', *Scientific American*, XV (2005), pp. 82–9.

5 J. Burke, 'Breaking the Wave', *Observer*, Sunday 27 February 2005.

6 Ibid.

7 The quotations are from ibid.

8 Tor Arne Andreassen, 'The Men Who Were Left Alone Again', *Aftenposten*, Sunday, 27 February 2005.

9 The World Bank, 'Rebuilding a Better Aceh and Nias: Stocktaking of the Reconstruction Effort', Brief of the Coordination Forum Aceh and Nias (CFAN) – October 2005, at http://siteresources.world bank.org/INTTSUNAMI/Resources/AcehReport9.pdf (2005).

10 Ibid.

11 The title of the article is 'When Earth Attacks', at www.popsci.com.

12 Quotation from a sermon given by Pastor Øyvind Valvik at the Holy Church of Jesus congregation in Kristiansand on 9 January 2005, at www.filakrs.no.

13 www.dagen.no: leader, Tuesday, 4 January 2005.

14 See the press release from Edvardsen at www.tbve.no.

15 Øystein Vik, 'The Tsunami is God's Punishment', www.dagen.no, Tuesday, 4 January 2005.

16 www.dunamai.com.

17 See for example K. A. Jacobsen, *Hinduism* (Oslo, 2003), for information about Hinduism.

18 For Buddhists and the tsunami, see Lisa Schneider, 'The Consolation of Karma', at www.beliefnet.com.

19 K. A. Jacobsen, *Buddhism* (Oslo, 2000).

20 Tor Arne Andreassen, 'Fears of Ghosts After the Tsunami', *Aftenposten*, 20 January 2005.

21 The Koran is available at www.muslimaccess.com, where this passage is taken from.

22 J. Homan, 'The Social Construction of Natural Disasters', in *Natural Disasters and Development in a Globalizing World*, ed. M. Pelling (London and New York, 2003), pp. 141–56.

23 T. Færøvik, *Veien til Xanadu. En reise i Marco Polos fotspor* (Cappelen, 2001).

24 Harun Yahya is a pseudonym used by Adnan Oktar. His enormous number of books gives the impression of having been written by a team of writers.

25 See www.harunyahya.com. A number of films and books dealing with the relation between natural disasters and Islam can be downloaded from the website.

26 Harun Yahya, 'The Truth of Life of This World', 3rd edn, at www.harunyahya.com (2002), p. 83.

27 Ibid., p. 80.

28 Ibid.

29 John Aglionby, 'Good Muslims Survived, Say Militants', *Guardian*, 8 January 2005.

30 The quotation is from a report in *Dagbladet* by Carsten Thomassen and Sveinung Uddu Ystad, Sunday, 9 January 2005, p. 14.

31 Tor Arne Andreassen, 'The Rebels Have a Base in Sweden', *Aftenposten*, Thursday, 3 February 2005.

32 John Aglionby, 'Legacy of Tsunami Brings Peace to Aceh', *Guardian*, 15 August 2005.

10 Perilous Nature

1 P. O'Keefe, K. Westgate and B. Wisner, 'Taking the Naturalness Out of Natural Disasters', *Nature*, CCLX (1976), pp. 566–7.

2 The information about the seminar is from correspondence with Kenneth Hewitt, Jean Copans and Paul Susman (2005/2006).

3 M. J. Watts, *Silent Violence: Food, Famine and Peasantry in Northern Nigeria* (Berkeley, CA, 1983).

4 K. Hewitt, 'The Idea of Calamity in a Technocratic Age', in *Interpretations of Calamity, The Risk and Hazard Series 1*, ed. K. Hewitt (Boston, MA, 1983), pp. 3–32 .

5 The reactions were not equally strong everywhere. In France, the book passed almost unnoticed because few people within the

social sciences were interested in environmental studies, and the eco-aware nature researchers were not particularly radical. (From correspondence with Jean Copans.)

6 Information about the historical development leading to disaster research is from D. Alexander, *Natural Disasters* (New York and London, 2001) and D. Chester, 'Volcanoes, Society and Culture', in *Volcanoes and the Environment*, ed. J. Marti and G. Ernst, pp. 404–39 (Cambridge, 2005).

7 K. Hewitt, *Regions of Risk: A Geographical Introduction to Disasters* (Harrow, 1997).

8 From personal correspondence with Hewitt in January 2006.

9 G. Bankoff, 'The Historical Geography of Disaster: "Vulnerability" and "Local Knowledge" in Western Discourse', in *Mapping Vulnerability: Disasters, Development and People*, edited by Bankoff, G. Frerks and D. Hilhorst (London and Sterling, VA, 2004), pp. 25–36.

10 Figures from www.em-dat.net. Disasters with fewer than ten victims are not included in the statistics.

11 See T. Steinberg, 'The Secret History of Natural Disaster', *Environmental Hazards*, III (2001), pp. 31–3, for a criticism of the lack of interest in natural disasters among historians.

12 B. McKibben, *The End of Nature: Humanity, Climate Change and the Natural World* (London, 2003), p. 83.

13 See for example R. Hooke, 'On the History of Humans as Geomorphic Agents', *Geology*, XXVIII (2000), pp. 843–6.

14 McKibben, *The End of Nature*, p. 90.

15 D. Chester, 'Theology and Disaster Studies: The Need for Dialogue', *Journal of Volcanology and Geothermal Research*, CXLVI (2005), pp. 319–28.

16 H. Schmuck, '"An Act of Allah": Religious Explanations for Floods in Bangladesh as Survival Strategy', *International Journal of Mass Emergencies and Disasters*, XVIII (2000), pp. 85–95.

17 S. M. Hoffman, 'After Atlas Shrugs: Cultural Change or Persistence after a Disaster', in *The Angry Earth: Disaster in Anthropological Perspective*, ed. A. Oliver-Smith and S. M. Hoffman (New York and London, 1999), p. 319.

18 Ibid.

19 Much material is available within this field. See for example S. M. Hoffman, 'The Worst of Times, the Best of Times: Towards a Model of Cultural Response to Disaster', in *The Angry Earth:*

Disaster in Anthropological Perspective, ed. A. Oliver-Smith and S. M. Hoffman (New York and London, 1999), pp. 134–55.

20 D. Keys, *Catastrophe* (New York, 1999).

21 J. Diamond, *Collapse: How Societies Choose to Fail or Survive* (London and New York, 2005).

22 See B. McGuire, 'In the Shadow of the Volcano', *Guardian*, 16 October 2003.

23 Quotations from B. McGuire, *A Guide to the End of the World: Everything You Never Wanted to Know* (Oxford, 2003).

24 M. Davis, *The Monster at Our Door: The Global Threat of Avian Flu* (New York and London, 2005).

25 www.islamonline.net. The comment was posted on 27 December 2004.

26 Comment on an Al Jazeera article.

27 www.islamonline.net. The comment was posted on 29 December 2004.

28 www.rawprint.com. Green was later acquitted in the Swedish high court.

Epilogue: Where the Devil Lives in the Ground

1 Our finds and the earlier theories about the fires are described in H. Svensen et al., 'Subsurface Fires in Mali: Refutation of the Active Volcanism Hypothesis in West Africa', *Geology*, XXXI (2003), pp. 581–4.

2 The disaster in the early 1970s is described by M. H. Glantz et al., *The Politics of Natural Disaster: The Case of the Sahel Drought* (New York, 1976).

3 See B. Fagan, *The Long Summer: How Climate Changed Civilization* (London, 2004), for more information about earlier climate in the Sahara.

4 http://www.wired.com/science/planetearth/magazine/16-06/ff_heresies_10worst.

BIBLIOGRAPHY

Adams, P. R. and G. R. Adams, 'Mount Saint Helens's Ashfall: Evidence
   for a Disaster Stress Reaction', *American Psychologist*, XXXIX (1984),
   pp. 252–60

Albala-Bertrand, J. M., *Political Economy of Large Natural Disasters* (Oxford,
   1993)

Alexander, D., *Natural Disasters* (New York and London, 2001)

—, 'Symbolic and Practical Interpretations of the Hurricane Katrina
   Disaster in New Orleans', at http://understandingkatrina.ssrc.org
   (2005)

Ambrose, S. H., 'Late Pleistocene Human Population Bottlenecks,
   Volcanic Winter, and Differentiation of Modern Humans', *Journal
   of Human Evolution*, XXXIV (1998), pp. 623–51

Amundsen, A. B., 'Konventikler og vekkelser', in *Norges religionshistorie*,
   ed. Bugge Amundsen (Universitetsforlaget, 2005), pp. 295–316

Anderson, A., *An Introduction to Pentecostalism* (Cambridge, 2004)

Arnold, D., 'Hunger in the Garden of Plenty: The Bengal Famine of
   1770', in *Dreadful Visitations: Confronting Natural Catastrophe in the Age
   of Enlightenment*, ed. A. Johns (New York, 1999), pp. 81–111

Bankoff, G., 'The Historical Geography of Disaster: "Vulnerability" and
   "Local Knowledge" in Western Discourse', in *Mapping Vulnerability:
   Disasters, Development and People*, edited by G. Bankoff, G. Frerks and
   D. Hilhorst (London and Sterling, VA, 2004), pp. 25–36

Baptista, M. A., J. M. Miranda, F. Chierici and N. Zitellini, 'New Study
   of the 1755 Earthquake Source Based on Multi-channel Seismic
   Survey Data and Tsunami Modeling', *Natural Hazards and Earth
   System Sciences*, III (2003), pp. 333–40

Barman, O., *Erindringer fra 1861 til 1867* (Trondhjem, 1904)

Bartleman, F., *Azusa Street* (New Kensington, PA, 1982)

Benedictow, O. J., *The Black Death 1346–1353: The Complete History* (Woodbridge, 2004)

Benjaminsen, T. A., and G. Berge, *Timbuktu. Myter, mennesker, miljø* (Oslo, 2000)

Berg, F. G. Steen, *Naturkatastrofer i Norge* (Bergen, 1937)

Bilham, R., 'A Flying Start, Then a Slow Slip', *Science*, CCCVIII (2005), pp. 1126–17

Birmingham, D., *A Concise History of Portugal* (Cambridge, 1993)

Black, G., *Triumph of the People. The Sandinista Revolution in Nicaragua* (London, 1981)

Bliksrud, L., G. Hestmark and T. Rasmussen, *Norsk Idéhistorie*, vol. IV: *Vitenskapens utfordringer* (Oslo, 2002)

Bloch-Hoell, N., *Pinsebevegelsen. En undersøkelse av pinsebevegelsens tilblivelse, utvikling og særpreg med særlig henblikk på bevegelsens utforming i Norge* (Oslo, 1956)

Blong, R. J., *Volcanic Hazards. A Sourcebook on the Effects of Eruptions* (Boston, MA, 1984)

Bondevik, S., J. I. Svendsen, G. Johnsen, J. Mangerud and P. E. Kaland, 'The Storegga Tsunami along the Norwegian Coast, its Age and Runup', *Boreas*, XXVI (1997), pp. 29–53

Borland, J., 'Stories of Ideal Japanese Subjects from the Great Kant Earthquake of 1923', *Japanese Studies*, XXV(2005), pp. 21–34

Brinkley, D., *The Great Deluge: Hurricane Katrina, New Orleans, and the Mississippi Gulf Coast* (New York, 2006)

Bruce, V., *No Apparent Danger: The True Story of Volcanic Disaster at Galeras and Nevado del Ruiz* (New York, 2001)

Burke, J., 'Breaking the Wave', *Observer*, Sunday 27 February 2005

Burroughs, W., ed., *Climate into the 21st Century* (Cambridge, 2003)

Carozzi, M., 'Reaction of British Colonies in America to the 1755 Lisbon Earthquake', *History of Geology*, II (1983), pp. 17–27

Chen, B., '"Resist the Earthquake and Rescue Ourselves": The Reconstruction of Tangshan After the 1976 Earthquake', I, *The Resilient City: How Modern Cities Recover from Disasters*, ed. L. J. Vale and T. J. Campanella (Oxford, 2005), pp. 235–53

Chester, D., 'The 1755 Lisbon Earthquake', *Progress in Physical Geography*, XXV (2001), pp. 363–83

—, 'Theology and Disaster Studies: The Need for Dialogue', *Journal of Volcanology and Geothermal Research*, CXLVI (2005), pp. 319–28.

—, *Volcanoes and Society* (London, Melbourne and Auckland, 1993)

—, 'Volcanoes, Society and Culture', in *Volcanoes and the Environment*, ed.
  J. Marti and G. G. J. Ernst (Cambridge, 2005), pp. 404–39

Cohn, N., *Noah's Flood: The Genesis Story in Western Thought* (New Haven
  CT, 1999)

—, *The Pursuit of the Millennium: Revolutionary Millenarians and Mystical
  Anarchists of the Middle Ages* (London, 1993)

Cyranoski, D., 'A Sleeping Giant Stirs', *Nature*, CDXXVIII (2004), pp. 12–13

Davies, R. W., and S. G. Wheatcroft, *The Years of Hunger: Soviet Agriculture,
  1931–1933* (Basingstoke, 2004)

Davis, M., *Dead Cities* (New York, 2002)

—, *Ecology of Fear: Los Angeles and the Imagination of Disaster* (Basingstoke
  and Oxford, 1998)

—, *Late Victorian Holocausts: El Niño Famines and the Making of the Third
  World* (London and New York, 2002)

—, *The Monster at Our Door: The Global Threat of Avian Flu* (New York and
  London, 2005)

—, 'The Predators of New Orleans', *Le Monde diplomatique*, October
  2005

De Oliveira Marques, A. H., *History of Portugal*, vol. I, *From Lusitana to
  Empire* (New York and London, 1972)

Diamond, J., *Collapse: How Societies Choose to Fail or Survive* (London and
  New York, 2005)

Doughty, P. L., 'Plan and Pattern in Reaction to Earthquake: Peru,
  1970–1998', in *The Angry Earth: Disaster in Anthropological Perspective*, ed.
  A. Oliver-Smith and S. L. Hoffman (New York and London, 1999),
  pp. 234–56

Dynes, R. R., 'The Dialogue Between Voltaire and Rousseau on the
  Lisbon Earthquake: The Emergence of a Social Science View',
  International Journal of Mass Emergencies and Disasters, XVIII (2000),
  pp. 97–115

Dyregrov, A., *Katastrofepsykologi* (Bergen, 2002)

Evans, R. J., 'Epidemics and Revolutions: Cholera in nineteenth-
  century Europe', in *Epidemics and Ideas: Essays on the Historical
  Perception of Pestilence*, ed. T. Ranger and P. Slack (Cambridge, 1992),
  pp. 158–63

Fagan, B., *Floods, Famines and Emperors: El Niño and the Fate of Civilizations*
  (New York, 1999)

—, *The Little Ice Age: How Climate Made History 1300–1850* (New York, 2002)

—, *The Long Summer: How Climate Changed Civilization* (London, 2004)

Favaro, A., C. Zaetta, G. Colombo and P. Santonastaso, 'Surviving the Vajont Disaster: Psychiatric Consequences 36 Years Later', *The Journal of Nervous and Mental Disease*, CXCII (2004), pp. 227–31

Fonseca, J. D., *1755: The Lisbon Earthquake* (Lisbon, 2005)

Fothergill, A., E.G.M. Maestas and J. D. Darlington, 'Race, Ethnicity and Disasters in the United States: A Review of the Literature', *Disaster*, XXIII (1999), pp. 156–73

Fulsås, N., *Havet, døden og vêret. Kulturell modernisering i Kyst-Noreg 1850–1950* (Oslo, 2003)

Furseth, A., 'Skredulykker i Norge', *Heimen*, XLI (2004), pp. 195–208

—, *Dommedagsfjellet: Tafjord 1934* (Gyldendal, 1994)

Færøvik, T., *Veien til Xanadu. En reise i Marco Polos fotspor* (Cappelen, 2001)

Galeano, E., *Memory of Fire*, III, *Century of the Wind* (New York, 1998)

Gilje, N., and T. Rasmussen, *Norsk Idéhistorie*, vol. II: *Tankeliv i den lutherske stat* (Oslo, 2002)

Glantz, M. H., et al., *The Politics of Natural Disaster: The Case of the Sahel Drought* (New York, 1976)

Godrej, D., *The No-Nonsense Guide to Climate Change* (London, 2001)

Goenjian, A. K., et al.: 'Posttraumatic Stress and Depressive Reactions Among Nicaraguan Adolescents after Hurricane Mitch', *American Journal of Psychiatry*, CLVIII (2001), pp. 788–94

Gottfried, R. S., *The Black Death: Natural and Human Disasters in Medieval Europe* (London, 1983)

Grattan, J., 'Pollution and Paradigms: Lessons from Icelandic Volcanism for Continental Flood Basalt Studies', *Lithos*, LXXIX (2005), pp. 343–53

Gunnarson, T., *Dommedagsventing. Millennismen og dens innslag i norsk kristendom* (Bergen, 1928)

Gunnarsson, Á., *Volcano: Ordeal by Fire in Iceland's Westmann Islands* (Rejkjavik, 1973)

Hallam, T., *Catastrophes and Lesser Calamities: The Causes of Mass Extinctions* (Oxford, 2004)

Hanska, J., 'Strategies of Sanity and Survival: Religious Responses to Natural Disasters in the Middle Ages', *Studia Fennica Historica*, II (Helsinki, 2002)

Harun Yahya, 'The Truth of Life of This World', 3rd edn, at www.harunyahya.com (2002)

Hedberg, B., *Kometer och kometskräck* (Stockholm, 1985, in Swedish)

Herbert, J., *Shintô. At the Fountain-Head of Japan* (New York, 1967)

Herstad, J., *I helstatens grep. Kornmonopolet 1735–88* (Oslo, 2000)

Hewitt, K., 'The Idea of Calamity in a Technocratic Age', in *Interpretations of Calamity, The Risk and Hazard Series 1*, ed. K. Hewitt, pp. 3–32 (London, 1983)

—, *Regions of Risk: A Geographical Introduction to Disasters* (Harrow, 1997)

Hodne, Ø., *Jutulhugg og riddersprang. Sagn fra norsk natur* (Oslo, 1990)

—, *Norsk folketro* (Oslo, 1999)

Hoffman, S. M., 'After Atlas Shrugs: Cultural Change or Persistence after a Disaster', in *The Angry Earth: Disaster in Anthropological Perspective*, ed. A. Oliver-Smith and S. M. Hoffman (New York and London, 1999), pp. 303–25

—, 'The Worst of Times, the Best of Times: Towards a Model of Cultural Response to Disaster', in *The Angry Earth. Disaster in Anthropological Perspective*, ed. A. Oliver-Smith and S. M. Hoffman (New York and London, 1999), pp. 134–55

Homan, J., 'The Social Construction of Natural Disasters', in *Natural Disasters and Development in a Globalizing World*, ed. M. Pelling (London and New York, 2003), pp. 141–56

Ingebrigtsen, G., I. Sandager, T. Sørensen and O. S. Dalgard, 'Når ulykken rammer. Posttraumatisk stressforstyrrelse i Norge, forekomst og relasjon til sosialt nettverk', I: *Sosialt nettverk, helse og samfunn*, ed. by O.S. Dalgard, E. Døhlie & M. Ystgaard (Universitetsforlaget, 1993), pp. 64–85

IPCC 2001, *Climate Change 2001: The Scientific Basis* (Cambridge, 2001)

IPCC 2007, *Climate Change 2007*, at www.ipcc.ch

Jacobsen, K. A., *Hinduism* (Oslo, 2003)

—, *Buddhism* (Oslo, 2000)

Jones, P. D., and K. R. Briffa, 'Unusual Climate in Northwest Europe during the Period 1730 to 1745 Based on Instrumental and Documentary Data', *Climate Change*, LXXIX, (2006), pp. 361–79

Justo, J. L., and C. Salwa, 'The 1531 Lisbon Earthquake', *Bulletin of the Seismological Society of America*, LXXXVIII (1998), pp. 319–28

Kates, R. W., et al., 'Human Impact of the Managua Earthquake', *Science*, CLXXXII, 7 December 1973, pp. 981–90.

Keane, S., *Disaster Movies: The Cinema of Catastrophe* (London and New York, 2001)

Kendrick, T. D., *The Lisbon Earthquake* (Philadelphia and New York, 1957)

Keys, D., *Catastrophe* (New York, 1999)

Nicolson, M. H., *Mountain Gloom and Mountain Glory: The Development of the Aesthetics of the Infinite* (Washington, DC, 1997)

Khoury, E. L., et al., 'The Impact of Hurricane Andrew on Deviant Behavior Among a Multi-Racial/Ethnic Sample of Adolescents in Dade County, Florida: A Longitudinal Analysis', *Journal of Traumatic Stress*, X (1997), pp. 71–91

Larsen, T., 'Ras, risiko og rasjonalitet – en studie av et skredtruet samfunn på Vestlandet', Masters thesis, University of Bergen (1998)

Lay, T., et al., 'The Great Sumatra-Andaman Earthquake of 26 December 2004', *Science*, CCCVIII (2005), pp. 1127–32

Lynas, M., *High Tide: News from a Warming World* (London, 2004)

McGuire, B., *A Guide to the End of the World: Everything You Never Wanted to Know* (Oxford, 2003)

—, 'In the Shadow of the Volcano', *Guardian*, 16 October 2003

McKibben, B., *The End of Nature: Humanity, Climate Change and the Natural World* (London, 2003)

Merchant, C., *Reinventing Eden: The Fate of Nature in Western Culture* (New York and London, 2004)

Morris, C., *The San Francisco Calamity by Earthquake and Fire* (Urbana, IL, 2002)

Munch, P. A., *Norrøne gude- og heltesagn* (Universitetsforlaget, 1996)

Møller, A., *Den farlege naturen* (Oslo, 1986)

Näsström, B-M., *Blot. Tro og offer i det førkristne Norden* (Oslo, 2001)

Nesdal, S., *Lodalen – fager og farleg* (Oslo, 2003)

Nesje, A., 'Ikkje gløymt etter 100 år.' geo, IV (2005), pp. 34–8

Nolan, M. L., 'Impact of Parícutin on Five Communities', in *Volcanic Activity and Human Ecology*, ed. P. Sheets and D. Grayson (Boston, MA, 1979), pp. 293–338

Næss, H. E., *Trolldomsprosessene i Norge på 1500–1600 tallet* (Universitetsforlaget, 1982)

Næss, L. O., and J. Vevatne, 'Klimatilpasning: lærdom fra tidligere flommer', *Cicerone*, II (2004)

O'Keefe, P., K. Westgate and B. Wisner, 'Taking the Naturalness Out of Natural Disasters', *Nature*, CCLX (1976), pp. 566–7

—, 'Peru's Five Hundred-Year Earthquake: Vulnerability in Historical Context', in *The Angry Earth: Disaster in Anthropological Perspective*, ed. A. Oliver-Smith and S. M. Hoffman (New York and London, 1999), pp. 74–88

—, 'Theorizing Disasters: Nature, Power, and Culture', in *Catastrophe and Culture: The Anthropology of Disasters*, ed. S. M. Hoffman and A. Oliver-Smith (Santa Fe, 2002), pp. 23–48

Oliver-Smith, A., and S. M. Hoffman, 'Why Anthropologists Should Study Disasters', in *Catastrophe and Culture: The Anthropology of Disasters*, ed. S. M. Hoffman and A. Oliver-Smith (Santa Fe, 2002), pp. 3–22.

Pelling, M., *The Vulnerability of Cities: Natural Disasters and Social Resilience* (London and Sterling, VA, 2003)

Pennock, R. T., *Tower of Babel: The Evidence Against the New Creationism* (Cambridge, MA, 2002)

Platt, R. H., *Disasters and Democracy: The Politics of Extreme Natural Events* (Washington, DC, 1999)

Pontoppidan, E., *Uforgribelige betænkninger over den naturlige aarsag til de mange og store jordskiælv, samt det usædvanlige veirlig, som nu paa nogen tid er fornummet, baade i og uden for Europa* (Kiöbenhavn, 1756)

Post, J. D., *Food Shortage, Climate Variability, and Epidemic Disease in Preindustrial Europe: The Mortality Peak in the Early 1740s* (Ithaca, NY, and London, 1985)

—, 'Nutritional Status and Mortality in Eighteenth-century Europe', in *Hunger in History: Food Shortage, Poverty, and Deprivation*, ed. L.F. Newman (Cambridge, MA, 1990), pp. 241–80

Quarantelli, E. L., ed., *What is a Disaster? Perspectives on the Question* (London and New York, 1998)

Reinhardt, O., and D. R. Oldroyd, 'Kant's Theory of Earthquakes and Volcanic Action', *Annals of Science*, XL (1983), pp. 247–72

Reinås, J., et al., '26.12. Rapport fra evalueringsutvalget for flodbølgekatastrofen i Sør-Asia', at www.evalueringsutvalget.no (2005)

Roald, L. A., 'Two Major Eighteenth-Century Flood Disasters in Norway', in *Palaeofloods, Historic Floods and Climate Variability: Applications in Flood Risk Assessment*, ed. V. R. Thorndycraft et al. (Proceedings of the PHEFRA workshop, Barcelona, October 2002, 2003), pp. 16–19.

Ruddiman, W. F., 'The Anthropogenic Greenhouse Era Began Thousands of Years Ago', *Climatic Change*, LXI (2003), pp. 261–93

Schmuck, H., '"An Act of Allah": Religious Explanations for Floods in Bangladesh as Survival Strategy', *International Journal of Mass Emergencies and Disasters*, XVIII (2000), pp. 85–95

Sen, A. K., *Poverty and Famines: An Essay on Entitlement and Deprivation* (Oxford, 1981)

Sigurdsson, H., and S. Carey, 'Volcanic Disasters in Latin America and the 13th November 1985 Eruption of Nevado del Ruiz Volcano in

Colombia', *Disasters*, x (1986), pp. 205–16

Stangeland, E., 'Å leve på en vulkan. Risiko og stedsidentitet på Vestmannaeyjar', Masters thesis, University of Oslo (2004)

Stein, R. S., 'Earthquake Conversations', *Scientific American*, xv (2005), pp. 82–9

Steinberg, T., *Acts of God: The Unnatural History of Natural Disasters in America* (Oxford, 2000)

—, 'The Secret History of Natural Disaster', *Environmental Hazards*, iii (2001), pp. 31–3

—, 'Smoke and Mirrors: The San Francisco Earthquake and Seismic Denial', in *American Disasters*, ed. S. Biel (New York, 2001), pp. 103–28

Steingrímsson, J., *Fires of the Earth* [1907] (Reykjavik, 1998)

Stensvold, A., 'Amerikansk vekkelseskristendom i Norge', i: *Norges religionshistorie*, ed. Bugge Amundsen (Universitetsforlaget, 2005), pp. 342–55

—, 'Kristen modernisering – misjon, indremisjon og frikirker', i: *Norges religionshistorie*, ed. Bugge Amundsen (Universitetsforlaget, 2005), pp. 317–41

Støylen, Ø., 'Å leve med skredfare. En studie om opplevelse av risiko og psykologiske belastninger', Masters thesis, University of Bergen (1999)

Stone, R., 'Iceland's Doomsday Scenario?', *Science*, cccvi (2004), pp. 1278–81

Strøm, H., *Physisk og Oeconomisk Beskrivelse over Fogderiet Søndmør, beliggende i Bergens Stift i Norge* (Kiøbenhavn, 1762–6)

Svendsen, G., and K. Werswick, *Fjellene dreper. Utsnitt av norsk katastrofehistorie gjennom 300 år* (Oslo, 1961)

Svensen, H., D. K. Dysthe, E. H. Bandlien, S. Sacko, H. Coulibaly, and S. Planke, 'Subsurface Fires in Mali: Refutation of the Active Volcanism Hypothesis in West Africa', *Geology*, xxxi (2003), pp. 581–4

—, D. A. Karlsen, A. Sturz, K. Backer-Owe, D. A. Banks and S. Planke, 'Processes Controlling Water and Hydrocarbon Composition in Seeps from the Salton Sea Geothermal System, California, usa', *Geology*, xxxv (2007), pp. 85–8

Søberg, O. R., 'En analyse av naturmotivene i Henrik Ibsens "Brand"', Masters thesis, University of Oslo (1995)

Terra Nova, 'Lisbon Recalled: All Saints Day, 1 November 1755', *Terra Nova*, iii (1991), pp. 670–72

Thorarinsson, S., 'Damage Caused by Volcanic Eruptions', i: *Volcanic*

Activity and Human Ecology, ed. P. D. Sheets and D. K. Grayson (Boston, MA, 1979), pp. 125–59

Vale, L. J., and T. J. Campanella, 'Axioms of Resilience', in *The Resilient City: How Modern Cities Recover from Disasters*, ed. T. J. Vale and L. J. Campanella (Oxford, 2005), pp. 335–55

—, 'The Cities Rise Again', in *The Resilient City: How Modern Cities Recover from Disasters*, ed. T. J. Vale and L. J. Campanella (Oxford, 2005), pp. 3–23

Voight, B., 'The 1985 Nevado del Ruiz Volcano Catastrophe: Anatomy and Retrospection', *Journal of Volcanology and Geothermal Research*, XLIV (1990), pp. 349–86

Weart, S. R., *The Discovery of Global Warming* (Cambridge and London, 2003)

White, L., 'The Historical Roots of our Ecological Crisis', *Science*, CLV (1967), pp. 1203–7

Wilkinson, D., *Silence on the Mountain: Stories of Terror, Betrayal, and Forgetting in Guatemala* (Boston and New York, 2001)

Williams, R. S., and J. G. Moore, 'Man Against Volcano. The Eruption of Heimaey, Vestmannaeyjer, Iceland', at www.ugs.gov (1983)

Williams, S. N., and S. Self, 'The October 1902 Plinian Eruption of Santa Maria Volcano, Guatemala', *Journal of Volcanology and Geothermal Research*, XVI (1983), pp. 33–56

Winchester, S., *Krakatoa: The Day the World Exploded, 27 August 1883* (London, 2004)

Wisner, B., P. Blaikie, T. Cannon, and I. Davis, *At Risk: Natural Hazards, People's Vulnerability and Disasters* (London and New York, 2004)

Witoszek, N., *Norske naturmytologier. Fra Edda til økofilosofi* (Oslo, 1998)

The World Bank, 'Rebuilding a Better Aceh and Nias: Stocktaking of the Reconstruction Effort', Brief of the Coordination Forum Aceh and Nias (CFAN) – October 2005, at http://siteresources.worldbank.org/INTTSUNAMI/Resources/AcehReport9.pdf (2005).

Zeilinga de Boer, J., and D. T. Sanders, *Earthquakes in Human History: The Far-reaching Effects of Seismic Disruptions* (Princeton, NJ, and Oxford, 2005)

—, *Volcanoes in Human History: The Far-reaching Effects of Major Eruptions* (Princeton, NJ, and Oxford, 2002)

ACKNOWLEDGEMENTS

A number of writers and researchers, thanks to the professional knowledge they have disseminated, have helped inspire me during work on this book. Many have also sown new ideas. I would like to single out Mike Davis's thought-provoking books as an important source of inspiration and knowledge, especially his *Late Victorian Holocausts*. Researchers such as Kenneth Hewitt, Ted Steinberg, David Chester, Jussi Hanska, Thomas Kendrick, Astor Furseth and Anthony Oliver-Smith, to name but a few, have all helped shape the content of this book through their respective specialities from the borderland between natural disasters and societies. In addition, a number of people have contributed with knowledge about various aspects of natural disasters and have made valuable contributions. *The End is Nigh* would not have been the same without discussions and comments (both 'in the field', during conversations and by e-mail) from Astor Furseth, Gabriela Garvalho, Eline Stangeland, Dan Banik, Erling Kvernevik, Kenneth Hewitt, Greg Bankoff, Anders Bødal, Paul Susman and Emanuel Minos. Several people have made useful comments on parts of the text, and contributed with excellent support underway. My thanks to Martine Bjørnhaug, Camilla Svensen, Bjørn Morten Litveit Hansen, Reidar Müller, Steinar Andreas Sæther and Albrecht Hofheinz. Thanks too to The Methodist Heritage Center in the USA for granting me access to Frank Bartleman's tract 'The Earthquake!!!' Furthermore, Marit Sørlie and Kristin Rangnes at the Geological Library of Oslo University have been extremely helpful in getting hold of the rapidly expanding volume of disaster literature. I would also like to thank Dag Kristian Dysthe, Einar H. Bandlien and Arne Dahr for good companionship during our expedition in Mali, and Sverre Planke from Volcanic Basin Petroleum Research, who made it possible. Finally, my thanks to chief editor Harald Engelstad at Aschehoug for his stimulating contributions and lasting support.

PHOTO ACKNOWLEDGEMENTS

The author and publishers wish to express their thanks to the below sources of illustrative material and/or permission to reproduce it.

Alberto Garcia: p. 156; Guntomara Art Presentation, Indonesia: p. 162; National Information Service for Earthquake Engineering, University of California, Berkeley: pp. 38-9, 60, 61, 118, 168; Chappatte's cartoon in NZZ am Sonntag (Zürich) courtesy of www.globecartoon.com: p. 80; photo courtesy of Astor Furseth: p. 94; Norges Geologiske Undersøkelse (Geological Survey of Norway): p. 109; www.atributetohinduism.com: p. 144; Cameron Reilly: p. 188.

INDEX